Proceedings of the NSAIS'16 Workshop on Adaptive and Intelligent Systems 2016

Organized by the North-European Society of Adaptive and Intelligent Systems

16.-17.8.2016 Lappeenranta, Finland

Editors: Mikael Collan & Pasi Luukka

LUT Scientific and expertise publications
Research Reports 58
ISSN 2243-3376

ISBN 978-952-265-986-6 (PDF)
ISBN 978-952-265-985-9

ISBN 9789522659866

90000 >

9 789522 659866

FOREWORD

This workshop NSAIS´16 is the second one in the series of workshops organized by the North-European Society for Adaptive and Intelligent Systems (NSAIS). The first workshop took place more than ten years ago and this is the first international event organized by NSAIS ever since. The idea of this workshop is to be the beginning of a series of workshops that take place typically bi-annually around the areas of interest to the NSAIS in Northern Europe.

This workshop is organized in association with the Lappeenranta University of Technology (LUT), the Finnish Operations Research Society (FORS), TRIZ Finland HUB, and the European Society for Fuzzy Logic and Technology (EUSFLAT) – the organizers thank the aforementioned organizations for their support.

This proceedings includes altogether seventeen papers, short papers, or abstracts of the submissions presented at the NSAIS´16 workshop that represent nine nationalities. The papers are a mix along the topical theme of the conference "OR+Fuzzy+TRIZ" with many contributions to decision-making. All papers have undergone peer review. The organizers thank the international scientific program committee and the reviewers for their support.

Pasi Luukka, Chairman, NSAIS´16 program committee, President of NSAIS

Mikael Collan, Co-Chair, NSAIS´16 program committee

NSAIS'16 International Scientific Program Committee

Pasi Luukka, LUT, Finland (Chair)
Mikael Collan, LUT, Finland, (Co-chair)
Mario Fedrizzi, U. Trento, Italy
Yuri Lawryshyn, U. Toronto, Canada
Christer Carlsson, Åbo Akademi, Finland
Jozsef Mezei, Åbo Akademi, Finland
Markku Heikkilä, Åbo Akademi, Finland
Leonid Chechurin, LUT, Finland
Olli Bräysy, U. Jyväskylä, Finland
Mariia Kozlova, LUT, Finland
Jan Stoklasa, LUT & Palacky, U., Finland & Czech Republic
Farhad Hassanzadeh, XPO Logistics, USA
Julian Yeomans, U. York, Canada
Matt Davison, U. Western Ontario, Canada
Jose Merigo Lindahl, Universidad de Chile, Chile
Matteo Brunelli, Aalto U., Finland
Jorma K. Mattila, LUT, Finland
Irina Georgescu, CSIE, Romania
Roberto Montemanni, IDSIA, Switzerland
Onesfole Kurama, LUT, Finland
Andri Riid, Tallinn UT., Estonia
Jyrki Savolainen, LUT, Finland
Azzurra Morreale, LUT, Finland
Iulian Nastac, Polytechnic U. Bucharest, Romania
Michele Fedrizzi, U. Trento, Italy
Jana Krejčí, U. Trento, Italy
Eeva Vilkkumaa, Aalto U., Finland

Local organizing committee

Pasi Luukka, LUT, Finland (Chair)
Mikael Collan, LUT, Finland (Co-Chair)

TABLE OF CONTENTS

Scientific papers

NSAIS16 Workshop Schedule

Tuesday – August 16th

<u>Venue: Skinnarilan Sauna</u>

09.00 Opening ceremony

09.15 Sessions I (Chair: Mikael Collan)

     1. Morreale, Stoklasa, and Talasek – Decision-making under different risk presentation forms: an empirical investigation from Czech Republic

     2. Luukka and Collan - Similarity based TOPSIS with WOWA aggregation

     3. Georgescu and Kinnunen - Mixed models for risk aversion, optimal saving and prudence

11.30 Lunch break (Lunch at the Student Union building)

12.20 Paper Sessions II (Chair: Jan Stoklasa)

     1. Kozlova, Luukka, Collan - Fuzzy inference system for real option valuation

     2. Stoklasa, Talasek, Kubatova, and Seitlova - Likert scales in group multiple-criteria evaluation: a distance-from-ideal based transition to fuzzy rules

     3. Lawryshyn, Luukka, Tam, Fan, and Collan - Estimating capital requirements for one-off operational risk events

13.45 Break

14.05 Paper Sessions II - (Chair: Jozsef Mezei)

     1. Brunelli and Mezei - Approximate operations on fuzzy numbers: are they reliable?

     2. Kukkurainen - Connection between triangular or trapezoidal fuzzy numbers and ideals

     3. Holecek and Talasova - Transitions between fuzzified aggregation operators

     4. Mattila - On Fuzzy-Valued Propositional Logic

15.45 End of the first workshop day

In the evening: Conference Evening program – Finnish Sauna with drinks and some BBQ

Wednesday – August 17th

<u>Venue: Auditorium 7343.1</u>

10.00 Keynote – Professor Christer Carlsson

11.00 Lunch break

12.00 Paper Session IV (Chair: Pasi Luukka)

> 1. Talasek and Stoklasa - Linguistic approximation under different distances/similarity measures for fuzzy numbers

> 2. De and Ahmed - A K-L divergence based Fuzzy No Reference Image Quality Assessment

> 3. Nastac, Dragan, Isaic-Maniu - Profitability analysis of small and medium enterprises in Romania using neural and econometric tools

> 4. Scherbacheva, Haario, and Killeen - Modeling host-seeking behavior of African malaria vector mosquitoes in presence of long-lasting insecticidal nets

13.40 Break

14.00 Paper Session V– (Chair: Leonid Chechurin)

> 1. Efimov-Soini, Kozlova, Luukka, and Collan - A multi-criteria decision-making tool with information redundancy treatment for design evaluation

> 2. Elfvengren, Lohtander, and Chechurin - Customer Need Based Concept Development

> 3. Kozlova, Chechurin, Efimov-Soini - Selecting an economic indicator for assisting theory of inventive problem solving

15.40 Closing ceremony

Decision-making under different risk presentation forms: an empirical investigation from Czech Republic

Azzurra Morreale
Lappeenranta University of Technology
School of Business and Management
Skinnarilankatu 32, 53851
Lappeenranta, Finland
Email: azzurra.morreale@lut.fi

Jan Stoklasa
Lappeenranta University of Technology
School of Business and Management
Skinnarilankatu 32, 53851
Lappeenranta, Finland
and
Palacký University, Faculty of Arts
Olomouc, Czech Republic
Email: jan.stoklasa@lut.fi

Tomáš Talášek
Lappeenranta University of Technology
School of Business and Management
Skinnarilankatu 32, 53851
Lappeenranta, Finland
and
Palacký University, Faculty of Arts
Olomouc, Czech Republic
Email: tomas.talasek@lut.fi

Abstract—The goal of this research is the empirical investigation of the effect of different forms of presentation of information concerning risky alternatives to people. Specifically, we further explore the findings of Morreale et al. from 2016, concerning the way how decision-makers use the information about possible future outcomes presented in various forms including histograms or continuous distributions in their decision-making. The results of a survey conducted on a sample of 26 bachelor students of Applied Economics in the Czech Republic provide evidence in favor of the existence of the irregularity in risk propensity connected with the amount and form of information presented to the decision-makers identified by Morreale et al. on an Italian sample. This previously undescribed change of risk propensity is associated with the mode of presentation of information, when the decision-maker is provided with the mean value of a risky alternative and the range of its possible values. Departure from this mode – i.e. either reduction or addition of information with respect to this setting can result in the decrease of risk propensity of the decision-maker.

I. INTRODUCTION

Even in their everyday lives people are expected to make difficult decisions objectively and rationally no matter how complex or uncertain the situation. In reality, decisions under uncertainty are often made based on intuition, leaving decision-making vulnerable to the pitfalls of cognitive biases. With the ever-increasing computational power, the outputs of sophisticated analytical and decision-support tools are usually available on-line, information is shared both in text and graphical form through social media, possible outcomes of the decision making situation are commonly visualized as histograms or as continuous distributions (probability density functions are provided to the decision makers in this case).

In this research, we study how decision-makers use the information about possible future outcomes presented in various forms, including histograms or continuous distributions in their decision-making. Our aim is to identify how the form and amount of presentation of information concerning risky events influences the decision making process and the propensity to risk-taking in decision makers.

This research follows the behavioral decision-making research perspective that posits human decision-making to be less than fully rational. Managers must often make decisions about complex strategic issues, like whether and how to proceed with business investments now and in the future, and they are expected to make them rationally and objectively. Generally, these decisions – that determine a firms success in a dynamic technological landscape – are made under uncertainty, either technological or market uncertainty. Yet, often managers make these decisions based on intuition and experience, risking that cognitive biases might influence the decision making process [13]. The classical theory of economic decision-making expects individuals to be fully rational. According to the assumptions of this theory, decision-makers would perceive similar levels of risk when faced with similar decision-making scenarios in the same context [10]. In reality, the form, in which the information relevant for the decision-making is presented to managers, can play a significant role in their risk perception. As a matter of fact, the manner in which people acquire knowledge about risk may influence the amount of risk they are willing to accept or the amount of risk they perceive [14], [6], which in turn affects their decisions.

These considerations raise the question of what is the best way to present information about risky investments. The focus of this paper is on how managers decision making and risk perception changes under different amount (and form) of information presented to them. The main research question we are addressing therefore is: *How does the form of presentation of information and the amount of information presented to the decision maker influence the economic decision making process?*

If the amount of information and the form in which it is presented play a significant role in managerial decision making, then attention has to be paid to these topics and potential pitfalls and biases identified. This is especially true when investors are facing investment decisions involving multiple possible outcomes. If much uncertainty is involved, such as in strategic decisions in technology-intensive industry, simulation methods can be used to provide insights into complex problems or to provide predictions. Distributions of possible outcomes represented by histograms, triangular and trapezoidal distributions or estimated continuous distributions, are common outputs presented to the managers as decision support [3], [4].

A common distribution of uncertain returns of R&D projects (e.g. pharmaceutical), is an asymmetrical distribution or, in other words, an overwhelming potential upside with little current downside exposure [7]. Such a way to communicate risk, may influence managers' risk taking [6]. Although most of the literature on decision making under uncertainty has concentrated on single numbers or simple gambles, very recently some papers [1], [14], [2], [6] have started investigating the importance of using graphical tools, e.g. distributions, in decision-making. However, whether using different ways to communicate risk – in particular, the transition from using single numbers to using continuous distributions – influences risk perception, has not been investigated. The only exception is represented in [8] where a specific irregularity in risk propensity connected with the amount and form of information presented to the decision-makers is identified. This irregularity occurs under the most common conditions that is, when the decision-maker is presented with a summary of the risky alternative consisting of its mean value and the range of its possible values. Either reduction of information (i.e. when the risky alternative is represented by just its mean value) or addition of information (i.e. when the risky alternative is represented by a histogram or a continuous distribution of the all possible returns) with respect to this setting may result in the decrease of risk propensity of the decision-maker. Statistically significant differences in the proportion of risk-averse answers were observed in the Italian sample as presented in [8].

This research follows this research stream. The paper focuses on the confirmation of the existence of this phenomenon, and explores its manifestation on a sample from a different country. Using a sample of 26 bachelor students of Applied Economics (Palacký University, Olomouc, Czech Republic) we have replicated the research presented in [8] in a different geographical context to find more evidence for (or evidence against) its manifestation under the specified circumstances.

II. METHODOLOGY, SAMPLE AND THE DESIGN OF THE SURVEY

This study examines students responses to a survey the aim of which was to capture how the amount of information provided influences their propensity to choose a risky alternative. Previously others have used similar methods to study related topics including risk-taking [12], [11], [9]. Specifically, 26 bachelor students of Applied Economics at Palacký University

Olomouc (Czech Republic, Faculty of Arts) participated in this research project on decision-making.

The survey concerned real-life decision-making and consisted of several items, where students had to choose between a sure gain (option A) and a risky one (option B), whose expected value was the same as in option A. The way to represent information under uncertainty (i.e. the risky option B) changed from using single numbers to using histograms or continuous distributions whose expected value was highlighted. All the items used very similar amounts of money and can thus be considered to be different representations of the same decision problem. To enhance the inter-item comparability of the respondents answers to items, we have set the standard deviation of the risky options to be equal to the mean value of the respective risky option in each item.

Although the survey itself comprised of more items, for the purpose of this paper we just introduce and present the results relative to four items relevant for our research question. These items will be referred to as ITEM (MEAN), ITEM (MEAN AND RANGE), ITEM (MEAN AND HIST) and ITEM (MEAN AND CONTINUOUS). Specifically, the risky option in ITEM (MEAN) was just represented by a single number, i.e. the average value. The risky option in ITEM (MEAN AND RANGE) was represented by the average value of a continuous distribution (in particular log-normal) and the "worst" and the "best" possible outcomes (in particular the 2.5^{th} and 97.5^{th} percentiles of the distribution), that somehow represent a measure of the variance of the distribution. As far as ITEM (MEAN AND HIST) and ITEM (MEAN AND CONTINUOUS) are concerned, the risky option was represented respectively by a histogram which approximates a lognormal distribution and by a lognormal distribution, both truncated at a preselected value for presentation purposes (the truncation was made always at the value of the 99th percentile of the respective distributions). We also showed the expected value for both the histogram and the continuous distribution.

In the following, the items we have provided to the students are shown in more details (see Figure 1). To check for order effects, that refer to the possibility that subjects' experience in a task might bias their decisions in following tasks, the items were presented to the students in a randomized order (6 versions of the questionnaire with a different order of items were used). For the purpose of this research we have constructed the items so that the amount of information contained within them would increase from ITEM (MEAN) through ITEM (MEAN AND RANGE) to ITEM (MEAN AND HIST) and also from ITEM (MEAN) through ITEM (MEAN AND RANGE) to ITEM (MEAN AND CONTINUOUS).

For each of the items, we asked the students to make their choice between option A and option B. The researcher informed the subjects that there was no right or wrong answer, and that responses were confidential and not connected in any way with the course evaluation (the survey was administered during one of the lectures).

In order to analyze our results (next section), we accept the definition of a risk-averse decision-maker according to

ITEM (MEAN)
- **A:** a certain gain of 27 EUR
- **B:** a lottery ticket to a lottery where the participant wins on average 27 EUR

ITEM (MEAN AND RANGE)
- **A:** a certain gain of 23 EUR
- **B:** a lottery ticket to a lottery where the participant wins on average 23 EUR, and the range of possible gains is from 3 EUR to 83 EUR

ITEM (MEAN AND HIST)
- **A:** a certain gain of 26 EUR
- **B:** a lottery ticket to a lottery whose possible gains are described by the following distribution. The average gain is 26 EUR

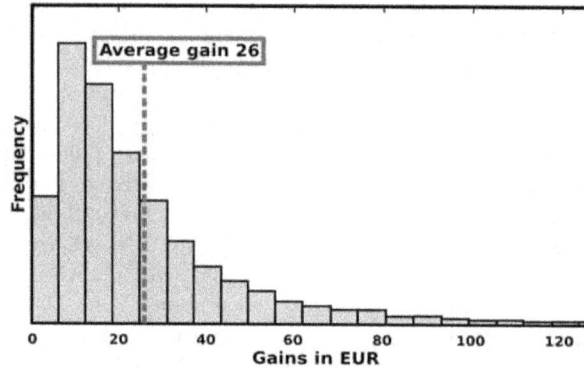

ITEM (MEAN AND CONTINUOUS)
- **A:** a certain gain of 24 EUR
- **B:** a lottery ticket to a lottery whose possible gains are described by the following distribution. The average gain is 24 EUR

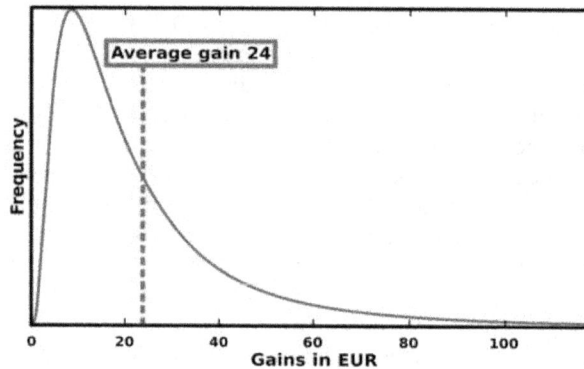

Fig. 1. A graphical summary of the four items provided to the student within the survey (English version – the actual survey was carried out in Czech, the sums of money were presented in CZK).

prospect theory [5] as applied also in [8], as an individual who, when faced with a choice between a certain outcome and a risky alternative with the same expected value, will choose the certain outcome.

III. RESULTS AND DISCUSSION

Even on the small sample of 26 bachelor students we have found evidence (though not statistically significant due to the smaller sample) of a violation of monotonicity of risk propensity with respect to the amount of information as described in [8]. The same pattern as the one identified in the Italian sample can be found in our data: increasing the amount of information from a simple expected value to a more information-rich form (expected value plus the worst and the best outcomes of the risky alternative) increases the propensity to take risk. On the contrary, further addition of information concerning the risky prospects seems to reduce the decision-makers propensity to risk. In fact, transition to a more complete information (graphical representation of the outcomes of a risky alternative) does not stimulate further the increase of the propensity to take risk. Figure 2 represents

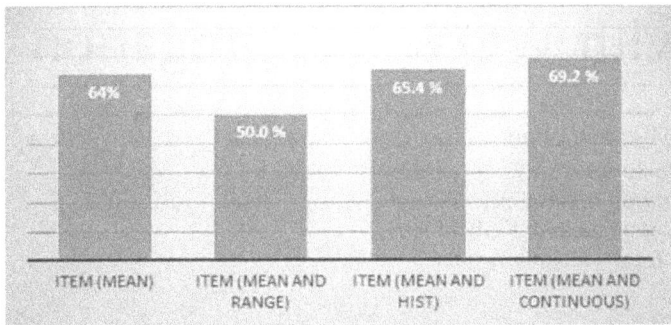

Fig. 2. Percentage of people who chose the safe option (A) over the four items

the percentage of people who chose the safe option (A) over the four afore described items. Although the results lack the statistical strength, the pattern is clearly visible in this data sample. Based on the data we can say that there seems to be no evidence against the phenomenon identified in [8].

We can observe (see Figure 2) that students were more risk adverse in ITEM (MEAN), i.e. when the risky option was represented by the lowest level of information (just the mean value). Adding the range of the distribution, i.e. the worst and the best outcomes (ITEM (MEAN AND RANGE)) increased risk taking. In fact, 50 percent of people (instead of 64%) chose the safe option. However, when information is definitely more complete as in ITEM (MEAN AND HIST) and ITEM (MEAN AND CONTINUOUS) than information in ITEM (MEAN AND RANGE), we observe that people are less likely to take risk and they behave as in ITEM (MEAN). Finally, we did not observe a big difference between results coming from ITEM (MEAN AND HIST) and ITEM (MEAN AND CONTINUOUS).

We have found evidence to support the claim that the propensity to choose risky alternatives does not change monotonically with the amount of information presented to the decision-maker. Although the size of the sample results in statistically insignificant differences between the different modes of information-presentation, the pattern from the Italian sample (where differences were statistically similar and the pattern analogical) repeated itself. The phenomenon therefore, in our opinion, seems to be present in the decision making under the specified conditions and deserves further investigation and will be the focus of our further research.

The fact that one of the most common means of presenting information concerning risky alternatives may also be causing a significant increase of risk propensity definitely deserves sufficient attention. Further investigation of this phenomenon may provide interesting impulse for the development of new economic decision-support tools and inspire further development of managerial decision support theory and practice. In addition, although this study focused only on positive outcomes (gains), future researcher will also focus on negative outcomes (losses) and investigate the risk propensity in such a domain.

IV. ACKNOWLEDGEMENTS

The research presented in this paper was partially supported by the grant IGA FF 2015 014 of the internal grant agency of Palacký University Olomouc.

REFERENCES

[1] S. Benartzi and R. Thaler, "Risk aversion or myopia? Choices in repeated gambles and retirement investments," *Management Sci.* vol. 45, no. 3, pp. 364-381, 1999.
[2] J. Beshears, J. Choi, D. Laibson and M. Madrian, "Can psychological aggregation manipulations affect portfolio risk-taking? Evidence from a framed field experiment." NBER Working Paper 16868, National Bureau of Economic Research, Cambridge, MA, 2011.
[3] M. Collan, *The Pay-Off Method: Re-Inventing Investment Analysis.* Charleston, NC, USA: CreateSpace, 2012.
[4] R. M. Hogarth and E. Soyer, "Communicating forecasts: The simplicity of simulated experience," *Journal of Business Research*, vol. 68, pp. 1800-1809, 2015.
[5] D. Kahneman and A. Tversky, "Prospect theory: An analysis of decision under risk," *Econometrica*, vol. 47, no. 2, pp. 263-292, 1979.
[6] C. Kaufmann, M. Weber and E. Haisley, "The Role of Experience Sampling and Graphical Displays on One's Investment Risk Appetite," *Management Science*, vol. 59, no. 2, pp. 323-340, 2013.
[7] G. Lo Nigro, A. Morreale, and L. Abbate, "An open innovation decision support system to select a biopharmaceutical R&D portfolio," *Managerial and Decision Economics*, 2015, DOI: 10.1002/mde.2727.
[8] A. Morreale, J. Stoklasa, M. Collan, G. Lo Nigro and T. Talášek, "The effect of multiple possible outcomes representation on managerial decision-making under uncertainty: an exploratory survey" in *Proc. 19th International Working Seminar on Production Economics*. Innsbruck, Austria, 2016.
[9] K. Miller and Z. Shapira, "An Empirical Test of Heuristics and Biases Affecting Real Option Valuation," *Strategic Management Journal*, vol. 25, pp. 269-284, 2004.
[10] P. C. Nutt, "Flexible decision styles and the choices of top executives," *Journal of Management Studies*, vol. 30, no. 5, pp. 695-721, 1993.
[11] M. Simon, S. M. Houghton and K. Aquino, "Cognitive biases, risk perception, and venture formation: How individuals decide to start companies," *Journal of Business Venturing*, vol. 14, no. 5, pp. 113-134, 2000.
[12] S. B. Sitkin and L. R. Weingart, "Determinants of risky decision-making behavior: A test of the mediating role of risk perceptions and propensity," *Academy of Management Journal*, vol. 38, no. 6, pp. 1573-1592, 1995.
[13] H. T. J. Smit and D. Lovallo, "Creating More Accurate Acquisition Valuations," *MIT Sloan Management Review*, vol. 56, pp. 63-72, 2014.
[14] E. U. Weber, N. Siebenmorgen and M. Weber. "Communicating asset risk: How name recognition and the format of historic volatility information affect risk perception and investment decision," *Risk Anal.*, vol. 25, no. 3, pp. 597-609, 2005.

Similarity based TOPSIS with WOWA aggregation

Pasi Luukka and Mikael Collan

School of Business
Lappeenranta University of Technology
P.O. Box 20, FIN-53851 Lappeenranta, Finland
E-mail: {pasi.luukka, mikael.collan}@lut.fi

Abstract—We propose a new Technique for Order Preference by Similarity to Ideal Solution (TOPSIS) variant which is based on similarity based TOPSIS and weighted ordered weighted averaging(WOWA). In generating weights for WOWA we apply the most often used regular increasing monotone (RIM) type quantifier. By doing this, we are able to create a generalized version of similarity based TOPSIS, of which the previously presented variants are sub-cases. By doing this generalization we also create an effect, where the ranking of alternatives can change depending on parameter value selection. For this purpose we apply the histogram ranking method that is able to take into consideration the variability of rankings and offers a robust holistic ranking of the alternatives. The proposed new method is applied to a patent portfolio selection problem.

Keywords—histogram ranking, similarity based TOPSIS, WOWA, RIM

I. INTRODUCTION

Technique for Order Preference by Similarity to Ideal Solution (TOPSIS) (see, e.g., [1]) has become a widely applied multi-criteria decision-making tool. The first TOPSIS variant extended to cover fuzzy numbers was introduced in the year 2000 by Chen [2]. Later on, fuzzy TOPSIS has been the subject of a wide variety of applications and further research that has created new variants (see, e.g., for reviews on the topic [3-5]).

One of the differences between the Ordered Weighted Average (OWA) [6] operator and the simple weighted mean is that OWA satisfies the commutativity condition. On the other hand, giving importance weights to, e.g., criteria to be aggregated, is often desirable. An operator that can combine these desired properties is called the Weighted Ordered Weighted Average (WOWA), and it was introduced in 1997 by Torra [7].

In this paper we generalize the earlier variants of similarity based TOPSIS [8] by aggregating similarities computed in similarity based TOPSIS by using the WOWA operator. The new variant proposed in this paper can produce the original method (see [8]) by setting equal weights for the both weighting vectors used. This allows one to examine how parameter values affect the ranking order (result) and how the to-be-compared alternatives are influenced by the parameter values. This information is often vital in order to make proper decisions and unfortunately often totally disregarded in many cases. This generalization ability of the new TOPSIS variant, unfortunately, comes at a cost: one must be able to properly and consistently deal with the needed parameters. For this purpose we apply the "histogram ranking" method [8] that can effectively offer help in gaining a more robust and complete understanding of the effect that parameter values have on the ranking of alternatives, when parametric methods are used. The histogram method offers an overall ranking that considers (full) ranges of parameter values.

The next section goes through the preliminaries and presents the "underlying concepts" of the new proposed method in detail. Then the new WOWA based similarity based TOPSIS variant is presented, followed by a numerical illustration of an application into a patent portfolio selection problem. The paper is closed with discussion and conclusions.

II. PRELIMINARIES

A. Similarity and weighted ordered weighted average

Term 'similarity' can be viewed from quite many different angles - Zadeh [9] gave the following definition for similarity in 1971:

Definition 1: A similarity relation, s, in X is a fuzzy relation in X, which is

 a) *reflexive, i.e. $s(x,x)=1$, $\forall x \in X$*
 b) *symmetric, i.e. $s(x,y)=s(y,x)$, $\forall x,y \in X$*
 c) *transitive, i.e. $s(x,z) \geq \sup_y (s(x,y)*s(y,z))$ $\forall x,y \in X$*

Transitivity in definition 1 is 'max-star' transitivity, but also other forms exist in the literature and there has been discussion about if the requirement of transitivity is necessary, e.g., see [10]. One commonly used similarity measure can be given as follows:

$$s(x,y)=1-|x-y| \qquad (1)$$

If we examine similarities s_i, $i = 1,...,n$ in a set X, we can define a binary relation by stipulating $S(x,y)=\frac{1}{n}\sum_{i=1}^{n} s_i(x,y)$, for all $x,y \in X$, see Turunen [11].

In the generalized version, similarity can be written as [12]:

$$s_p(x,y) = (1 - |x^p - y^p|)^{1/p} \qquad (2)$$

In the TOPSIS, one examines a choice-situation, where one has multiple criteria for each alternative and the observed value of each criterion is compared to a negative and a positive ideal solution. The comparison can be conducted by using a similarity (and not a distance). This creates a need for computing several similarities between a sample and an ideal solution and aggregating these to a single similarity value. For this purpose and instead of using a standard arithmetic mean, we use a more general aggregation operator, the WOWA [7].

Definition 2: Let z and w be n dimensional normalized weighting vectors and Q be a Regular Increasing Monotonic quantifier such that $Q(0)=0$ and $Q(1)=1$. A mapping $S_{WOWA}:[0,1]^n \rightarrow [0,1]$ is WOWA operator of n dimension with respect to w and Q if:

$$S_{WOWA}(s_1, s_2, \ldots, s_n) = \sum_{i=1}^{n} w_i s_{\pi(i)} \qquad (3)$$

where $s_{\pi(i)}$ is the ith largest of elements $(s_1, s_2, \ldots, s_n)$, and the weight w_i is defined as:

$$w_i = Q\left(\sum_{j=1}^{i} z_{\pi(j)}\right) - Q\left(\sum_{j=1}^{i-1} z_{\pi(j)}\right), \quad 1 \leq j \leq n \qquad (4)$$

Basic RIM quantifier Q for weight generation is $Q(r) = r^m$, $m > 0$.

In the following section we present the resulting new variant for TOPSIS.

III. WOWA BASED SIMILARITY BASED TOPSIS

Next we introduce a new similarity based TOPSIS variant that is based on WOWA based similarity. The original TOPSIS, see, e.g., [1], is based on simultaneously considering the distance from a studied "alternative" to a positive and to a negative ideal solution; these distances are calculated for each studied alternative and the TOPSIS method selects as the best alternative the alternative that has the most relative closeness to the positive ideal solution and that has the longest relative distance from the negative ideal solution. This idea has been extended, and one variant of TOPSIS uses the concept of similarity to replace the distance used in the original TOPSIS [8]. In this paper we further extend the similarity based TOPSIS by further generalizing the aggregation operator used in the calculation of the similarity vector by using a WOWA operator in the aggregation.

The procedure of using the proposed method starts from the construction of an evaluation matrix $\tilde{X} = [x_{ij}]$. Where x_{ij} denotes the score of the ith alternative, with respect to the jth criterion. The rest of the procedure can be summarized in six steps as follows:

Step 1: Normalization of the decision matrix $R=[r_{ij}]$

$$r_{ij} = \frac{x_{ij} + \left|\min_i(x_{ij})\right|}{\max_i(x_{ij}) - \min_i(x_{ij})}, i = 1, \ldots, l, j = 1, \ldots, n \qquad (5)$$

Step 2: Construction of the weighted normalized decision matrix $V=[v_{ij}]$

$$v_{ij} = r_{ij}(\cdot)w_j, i = 1, \ldots, l, j = 1, \ldots, n \qquad (6)$$

Step 3: Determination of the positive and the negative ideal solutions A^+ and A^-:

$$A^+ = \{v_1^+, \ldots, v_m^+\} = \{(max_j v_{ij}|j \in B), (min_j v_{ij}|j \in C)\} \quad (7)$$
$$A^- = \{v_1^-, \ldots, v_m^-\} = \{(min_j v_{ij}|j \in B), (max_j v_{ij}|j \in C)\}$$

where B stands for benefit criteria, and C for cost criteria.

Step 4: Calculation of the similarity of each alternative to the positive ideal solution and to the negative ideal solution:

$$s_i^+ = \sum_{k=1}^{n} w_k s_{\pi(k)}^+, i=1, \ldots, l \qquad (8)$$

where $s_{\pi(k)}^+$ is the kth largest of elements $(s_1^+, s_2^+, \ldots, s_n^+)$, where individual similarity is computed using equation (2) and the weight w_k is defined according to equation (4).

$$s_i^- = \sum_{k=1}^{n} w_k s_{\pi(k)}^-, i=1, \ldots, l \qquad (9)$$

where $s_{\pi(k)}^-$ is the kth largest of elements $(s_1^-, s_2^-, \ldots, s_n^-)$, where individual similarity is computed using equation (2) and the weight w_k is defined according to equation (4).

Step 5: Calculation of the relative closeness to both the ideal solutions (closeness coefficient):

$$CC_i = \frac{s_i^+}{s_i^+ + s_i^-}, i=1, \ldots, l \qquad (10)$$

Step 6: Ranking the alternatives: the closer the CC_i is to one, the higher the priority of the i:th alternative is.

A. Histogram ranking with similarity based TOPSIS

The histogram ranking method [8] solves the problem of parameter value selection by forming a "parameter value independent" ranking of alternatives. In order to apply the method, first a ranking of the alternatives is performed with a range of discrete parameter values that range from the predefined boundaries (i.e., from min to max) – this means that as many ranking instances are done as there are discrete parameter values used. Then, for each alternative a histogram is created from the received rankings. A center of gravity (COG) for each "ranking histogram" is calculated. The COG for each histogram is calculated as:

$$\bar{x}_j = \frac{\sum_{i=1}^n x_i H_i}{\sum_{i=1}^n H_i}, \qquad \bar{y}_j = \frac{\frac{1}{2}\sum_{i=1}^n H_i^2}{\sum_{i=1}^n H_i} \qquad (11)$$

where x_i represents the ranking order i and H_i frequency of this particular ranking order for the j:th alternative. The final ranking of alternatives is done by using the center of gravity point $\bar{x}_i$ and ranking the alternatives to ascending order according to this value. If the $\bar{x}_i$ is equal for two or more alternatives, then the distance of the center of gravity point $(\bar{x}_i, \bar{y}_i)$ from the Origo is used to determine the ranking order of the alternatives with the same rank. For more information about the method see [8]. This method is applied here to the parameter dependent closeness coefficients used in the introduced version of the TOPSIS method.

IV. WOWA BASED SIMILARITY BASED TOPSIS APPLIED TO A PATENT SELECTION PROBLEM

The problem presented here deals with ranking of patents and the selection of the best ranking patents to be included in a patent portfolio, this problem has been previously presented in [13]. The initial (financial) evaluation of the patents has been done by a group of managers and has resulted in a consensual fuzzy pay-off distribution for each patent. The possibilistic moments of these pay-off distributions (that are fuzzy numbers) are used as the criteria for ranking of the patents and are shown in Table 1.

Table 1: Possibilistic moments from the consensus pay-off distributions for each patent [13]

Patent	Mean	Standard deviation	Skewness
1	0.3175	0.0081	0.0005
2	0.3593	0.0111	0.0011
3	0.3203	0.0044	-0.0001
4	0.3038	0.0081	-0.0004
5	0.2665	0.0010	0.0000
6	0.4546	0.0174	-0.0018
7	0.4447	0.0143	0.0014
8	0.3504	0.0092	0.0013
9	0.3301	0.0083	0.0004
10	0.3297	0.0022	0.0002
11	0.3178	0.0067	0.0012
12	0.3638	0.0037	0.0001
13	0.2900	0.0038	0.0002
14	0.3352	0.0049	0.0001
15	0.3536	0.0069	0.0008
16	0.4187	0.0127	0.0011
17	0.5409	0.0169	0.0003
18	0.3934	0.0126	0.0012
19	0.4125	0.0111	-0.0001
20	0.3042	0.0050	0.0005

In Table 2 we show the closeness coefficient values and resulting rankings with. two different p parameter values.

Table 2: Ranking of the patents with two parameter p values

Patent	CC	Ranking order	CC	Ranking order

		(p=1)		(p=5)
1	0.4393	12	0.8969	10
2	0.2964	16	0.7497	15
3	0.6009	3	0.9595	3
4	0.5622	6	0.9597	2
5	0.6836	1	0.9676	1
6	0.5076	9	0.4954	18
7	0.1738	20	0.3527	19
8	0.3132	15	0.7512	14
9	0.4465	11	0.9044	9
10	0.6156	2	0.9420	6
11	0.3941	14	0.8473	13
12	0.5866	4	0.9352	7
13	0.5826	5	0.9506	4
14	0.5618	7	0.9448	5
15	0.4239	13	0.8575	12
16	0.2504	17	0.6163	17
17	0.2491	18	0.2324	20
18	0.2450	19	0.6261	16
19	0.4526	10	0.8650	11
20	0.5136	8	0.9187	8

It is clearly shown that the strength of the similarity value (defined by parameter p) influences the ranking.

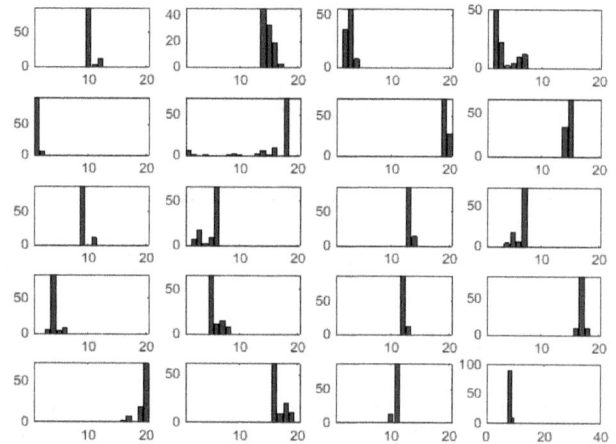

Figure 1: Histograms of the cumulative rankings of the twenty alternatives with a hundred different parameter p values; $p=0.1:0.1:10$ (between 0.1 and 10, with a step of 0.1), parameter m is set to $m=1$ in the computations.

The closeness coefficient values´ changes for each alternative patent, while the parameter value p is changed is visualized in Figure 2.

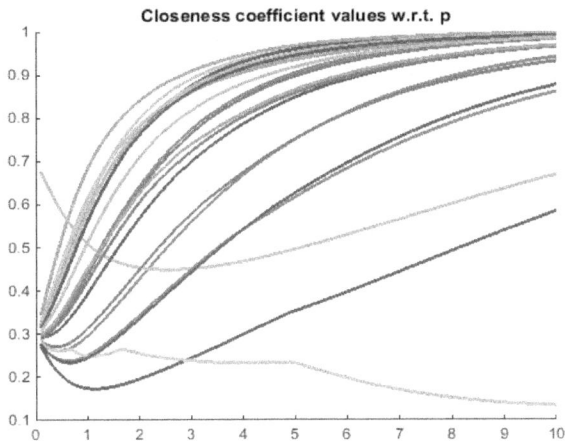

Figure 2: Closeness coefficient values, when changing the parameter p value for each patent.

Next, we examine more thoroughly the results shown in the histograms presented in Figure 1 by applying the histogram ranking method. The cumulative ranking histograms is an intuitively understandable way to show the "weight" of different rankings that each alternative gets, when parameter values of the basis method change. Results for the twenty patents gained from using the histogram ranking method are given in Table 3.

Table 3: Center of gravity points for each patent, when parameter p is changed

Patent	$\bar{x}$	$\bar{y}$	Ranking order
1	10.26	37.72	10
2	14.77	17.85	15
3	2.72	22.48	2
4	3.34	16.65	3
5	1.06	44.36	1
6	15.77	25.33	16
7	19.28	29.84	19
8	14.66	27.56	14
9	9.22	40.21	9
10	5.12	23.83	5
11	13.14	37.96	13
12	6.48	28.35	7
13	4.15	34.97	4
14	5.65	23.83	6
15	12.12	39.44	12
16	17	34.43	18
17	19.51	27.81	20
18	16.79	21.51	17
19	10.87	38.69	11
20	8.09	41.81	8

The best five patents ranked in this way are 5, 3, 4, 13, and 10. Next, we examine the overall effect of the "quantifier parameter" m to see how it affects the overall ranking results.

In Figure 3, one can see the produced histograms, when the values for the parameter m are changed and in Figure 4 one can see how closeness coefficient values for each patent change, when we change the parameter m value.

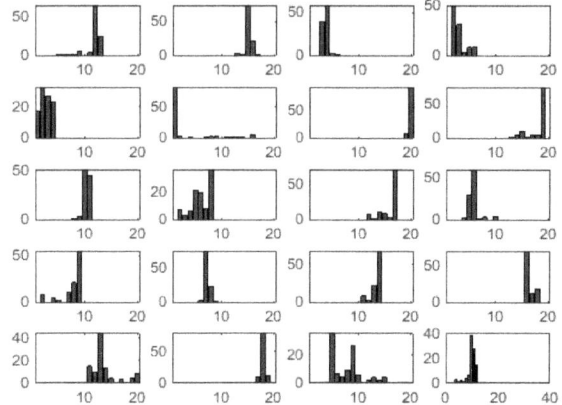

Figure 3: Histograms of the cumulative rankings of the twenty alternatives with a hundred different parameter m values; m=0.1:0.1:10 (between 0.1 and 10, with a step of 0.1). Parameter p is now set to p=1.

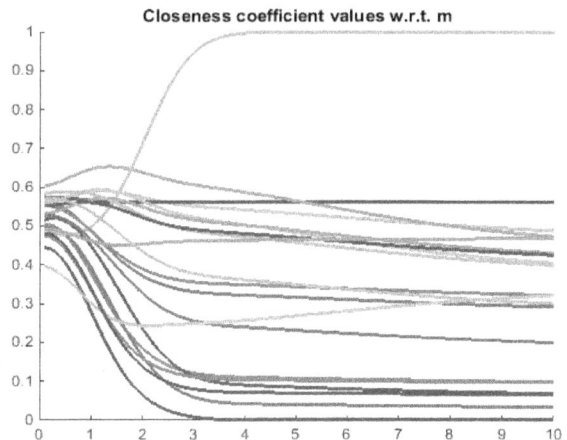

Figure 4: Closeness coefficient values when changing the parameter m value for each patent.

In Table 4, we compute the center of gravity points and the ranking order for the histograms created by varying the value of parameter m. From this experiment we get a result that suggests that the top five patents are patents 5,6,4,3 and 12.

Table 4: Center of gravity points for each patent, when parameter m is varied.

Patent	$\bar{x}$	$\bar{y}$	Ranking order
1	11.84	23.55	12
2	15.16	29.64	15
3	3.65	24.45	4
4	2.93	17.99	3

5	2.56	13.18	1
6	2.62	13.18	2
7	19.92	42.64	20
8	18.19	29.63	19
9	10.40	23.18	11
10	6.10	11.28	6
11	16.09	25.72	16
12	5.89	22.65	5
13	7.75	18.25	8
14	7.23	30.05	7
15	13.50	25.88	13
16	16.48	26.84	17
17	13.74	12.58	14
18	18.02	31.64	18
19	7.84	10.88	9
20	10.09	13.05	10

It can be seen that the parameter values p and m clearly affect the ranking of the patents.

V. DISCUSSION AND CONCLUSIONS

In this paper we have proposed a WOWA based similarity based TOPSIS method. We applied the basic RIM quantifier to generate weights in WOWA. When we applied the general form of the similarity, we noticed that ranking order of patents, when ranked using WOWA based similarity based TOPSIS altered depending on p parameter value. To investigate this and to get to final ranking order, we applied the histogram ranking method to get an overall ranking that considers the effect of parameter value changes.

A numerical illustration was made to show how using histogram ranking we can generate ranking orders which can take into account the different rankings gained by individual rankings when using a single parameter value. This kind of examination of the ranking results is vital, since patents that are consistently performing well can be more easily chosen (like, e.g., patent no 5), and patents that have a larger variety of ranking results (like, e.g., patent no 19 in Figure 3, or patent number 6 in Figure 1) can be left out. This kind of study shows that it is possible to get high ranking results for patents with single parameter values, but that the same high-ranking patents can after a "parameter study" be seen to have high uncertainty in consistency of the ranking results - one should carefully consider the selection of such inconsistent patents.

[1] C.-L. Hwang, K. Yoon, Multiple Attributes Decision Making Methods and Applications, Springer, Berlin, Heidelberg, 1981

[2] C. T. Chen, "Extensions of the TOPSIS for group decision-making under fuzzy environment," *Fuzzy Sets Syst.*, vol. 114, pp. 1–9, 2000.

[3] C. Carlsson, R. Fuller, "Fuzzy multiple criteria decision making: Recent developments", Fuzzy Sets and Systems 78, 1996, pp. 139-153.

[4] R.A. Ribeiro, "Fuzzy Multiple Attribute Decision Making: A Review and New Preference Elicitation Techniques", Fuzzy Sets and Systems, 78, 1996, pp. 155-181.

[5] C. Kahraman, S.C. Onar, B. Oztaysi, "Fuzzy Multicriteria Decision-Making: A Literature Review", International Journal of Computational Intelligence Systems, 8(4), 2015, pp. 637-666.

[6] R.R. Yager, "On ordered weighted averaging aggregation operators in multi-criteria decision making", IEEE Transactions on Systems, Man, and Cybernetics, 18, 1988, pp. 183-190.

[7] V. Torra, "The Weighted OWA operator", International Journal of Intelligent Systems, 12, 1997, pp. 153-166.

[8] P. Luukka, M. Collan, Histogram ranking with generalized similarity based TOPSIS applied to patent ranking. International Journal of Operational Research, 25(4), (2016), pp. 437-448.

[9] L. A. Zadeh, "Similarity relations and fuzzy orderings," *Information Sciences*, vol. 3, pp. 177-200, 1971.

[10] F. Klawonn, 2003, "Should fuzzy equality and similarity satisfy transitivity?", Fuzzy Sets and Systems, 133(2), 175–180.

[11] E. Turunen, Mathematics behind fuzzy logic, Physica Verlag, Heidelberg

[12] P. Luukka, K. Saastamoinen, V. Könönen, "A classifier based on the maximal fuzzy similarity in the generalized Łukasiewicz structure", Proc of the FUZZ-IEEE 2001 Conference, 2.-5.12.2001, Melbourne, Australia, 195-198

[13] M. Collan, M. Fedrizzi, P. Luukka, "A multi-expert system for ranking patents: An approach based on fuzzy pay-off distributions and a TOPSIS–AHP framework", Expert Systems with Applications, 40(12), 2013, pp. 4749–4759.

Mixed Models for Risk Aversion

Irina Georgescu

Dept. of Economic Information and Cybernetics
Academy of Economic Studies
Bucharest, Romania
irina.georgescu@csie.ase.ro

Jani Kinnunen

Åbo Akademi, IAMSR
Turku, Finland
jpkinnunen@gmail.com

Abstract— The models of this paper refer to mixed risk situations: one parameter is a fuzzy number, and the other is a random variable. Three notions of mixed expected utility are proposed as a mathematical basis of these models. The results of the paper describe the risk aversion of an agent in front of a risk situation with mixed parameters.

Keywords—mixed expected utility; possibilistically risk averse agent; possibilistically risk averse agent; mixed risk averse agent

I POSSIBILISTIC INDICATORS OF FUZZY NUMBERS

In this section we will recall two possibilistic indicators of fuzzy numbers: expected value and variance (according to [1], [2], [3], [4], [5], [6]).

Definition 1 A fuzzy subset A of $\mathbf{R}$ is a *fuzzy number* if it is upper semi-continuous, normal, fuzzy convex and supp(A) is a bounded subset of $\mathbf{R}$.

A function f:[0,1]→$\mathbf{R}$ is a *weighting function* if it is non-negative, monotone increasing and satisfies the normalization condition $\int_0^1 f(\gamma)d\gamma = 1$.

We fix a weighting function f:[0,1]→$\mathbf{R}$ and a fuzzy number A whose level sets are $[A]^\gamma = [a_1(\gamma), a_2(\gamma)]$ for $\gamma \in [0,1]$.

The *f-weighted possibilistic expected value* E(f, A) of the fuzzy number A is introduced by

$$E(f, A) = \frac{1}{2}\int_0^1 (a_1(\gamma) + a_2(\gamma))f(\gamma)d\gamma \quad (1)$$

When f(γ)=2γ, $\gamma \in [0,1]$, this notion was defined in [1]; in form (1) it has been introduced in [3].

There are several proposed notions of f-weighted possibilistic variance of a fuzzy number A. We will present three of them, keeping the notations of [5]:

$$Var_1(f, A) = \frac{1}{12}\int_0^1 [a_2(\gamma) - a_1(\gamma)]^2 f(\gamma)d\gamma \quad (2)$$

$$Var_2(f, A) = \frac{1}{2}\int_0^1 ([a_1(\gamma) - E(f, A)]^2 + [a_2(\gamma) - E(f, A)]^2)f(\gamma)d\gamma \quad (3)$$

$$Var_3(f, A) = \int_0^1 [\frac{1}{a_2(\gamma) - a_1(\gamma)}\int_{a_1(\gamma)}^{a_2(\gamma)} (x - E(f, A))^2 dx]f(\gamma)d\gamma \quad (4)$$

When f(γ)=2γ, $\gamma \in [0,1]$, the first two possibilistic covariances have been defined in [1]. The form (2) appears in [3] and (3) appears in [6]. The variance $Var_3(f, A)$ has been introduced [4] to evaluate the possibilistic risk premium. According to [5], Proposition 3.4.7, there exists the following relation between the three possibilistic variances:

$$Var_1(f, A) = \frac{1}{2}(Var_2(f, A) - Var_3(f, A)) \quad (5)$$

II THREE NOTIONS OF MIXED EXPECTED UTILITY

In this section we will study three notions of mixed expected utility associated with a mixed vector and a bidimensional utility function. They will correspond to the three possibilistic variances of the previous section.

Mixed vectors model uncertainty situations with two risk parameters: one is described by a fuzzy number, and the other by a random variable.

All the random variables of this paper will be defined on a fixed probability space (Ω, Κ, P). The expected value of a random variable X will be denoted by M(X), and its variance by Var(X). If g:$\mathbf{R}$→$\mathbf{R}$ is a unidimensional utility function, then g(X) is a random variable and the expected value M(g(X)) will be called probabilistic expected utility of X w.r.t. g.

We fix a weighting function f:[0,1]→$\mathbf{R}$ and a bidimensional utility function u:$\mathbf{R}^2$→$\mathbf{R}$ of class C^2.

We consider *a mixed vector* (A, X) in which A is a fuzzy number, and X is a random variable. Assume that the level sets of A are $[A]^\gamma = [a_1(\gamma), a_2(\gamma)]$ for $\gamma \in [0,1]$.

For any a∈$\mathbf{R}$, we consider a random variable u(a, X):Ω→$\mathbf{R}$ defined by u(a, X)(ω)=u(a, X(ω)) for any ω∈Ω.

Definition 2 We define the following *three mixed expected utilities* associated with f, the utility function g and the mixed vector (A, X):

$$E_2(f, u(A, X)) = \frac{1}{2}\int_0^1 [M(u(a_1(\gamma), X)) + M(u(a_2(\gamma), X))]f(\gamma)d\gamma \quad (6)$$

$$E_3(f, u(A, X)) = \int_0^1 [\frac{1}{a_2(\gamma) - a_1(\gamma)} \int_{a_1(\gamma)}^{a_2(\gamma)} M(u(y, X))dy]f(\gamma)d\gamma$$

$$\tag{7}$$

$$E_1(f, u(A, X)) = \frac{1}{2}[E_2(f, u(A, X)) - E_3(f, u(A, X))] \tag{8}$$

Remark 1 $E_2(f, u(A, X))$ has been introduced in [5], Chapter 7. The new expected utility $E_3(f, u(A, X))$ extends the unidimensional expected utility from [4] Definition 4.2.1 and the form of $E_1(f, u(A, X))$ is inspired by relation (5) of Section 1.

Lemma 1 (i) If the fuzzy number A is the fuzzy point $\bar{a}$ with $a \in \mathbf{R}$ then:

$$E_j(f, u(A, X)) = M(u(a, X)) \text{ for } j=1,2,3$$

(ii) If the random variable X has the form $\begin{pmatrix} b \\ 1 \end{pmatrix}$ with $b \in \mathbf{R}$ then:

$$E_2(f, u(A, X)) = \frac{1}{2}\int_0^1 [u(a_1(\gamma), b) + u(a_2(\gamma), b)]f(\gamma)d\gamma$$

$$E_3(f, u(A, X)) = \int_0^1 [\frac{1}{a_2(\gamma) - a_1(\gamma)} \int_{a_1(\gamma)}^{a_2(\gamma)} u(y, b)dy]f(\gamma)d\gamma$$

(iii) If A is the fuzzy point $\bar{a}$ and X is $\begin{pmatrix} b \\ 1 \end{pmatrix}$ then:

$$E_j(f, u(A, X)) = u(a, b) \text{ for } j=1,2,3$$

Each of the three notions of Definition 1 can be the starting point of a "mixed expected utility theory" (=MEU theory), in whose framework to be able to develop a risk theory with mixed parameters.

The following propositions lay stress on mathematical properties which show that the notions of Definition 1 are able to fulfill this goal.

Proposition 1 Let g, h be two bidimensional utility functions and $\alpha, \beta \in \mathbf{R}$. If $u = \alpha g + \beta h$ then:

(i) $E_j(f, u(A, X)) = \alpha E_j(f, g(A, X)) + \beta E_j(f, h(A, X))$ for j=1,2,3

(ii) $g \leq h$ implies $E_j(f, g(A, X)) \leq E_j(f, h(A, X))$ for j=1,2,3.

Proposition 2 If $u(y, x) = (y - E(f, A))(x - M(X))$ for any x, $y \in \mathbf{R}$, then $E_j(f, u(A, X)) = 0$ for any j=1,2,3.

Next we use the usual notations:

$$u_1 = \frac{\partial u}{\partial y}, u_2 = \frac{\partial u}{\partial x}, u_{11} = \frac{\partial^2 u}{\partial y^2}, u_{22} = \frac{\partial^2 u}{\partial x^2}, u_{12} = u_{21} = \frac{\partial^2 u}{\partial y \partial x},$$

etc.

Proposition 3 For any $j \in \{1,2,3\}$ the following approximation formula holds:

$$E_j(f, u(A, X)) \approx u(E(f, A), M(X)) +$$

$$\frac{1}{2}u_{11}(E(f, A), M(X))Var_j(f, A) +$$

$$+ \frac{1}{2}u_{22}(E(f, A), M(X))Var(X)$$

III CONVEXITY CONDITIONS

In this section we will prove some results which connect convexity and concavity of a utility function u(y, x) with respect to each of y and x. They will be applied in the next section to discuss some topics of mixed risk.
We fix a weighting function $f:[0, 1] \to \mathbf{R}$.
We recall next three classic results on the convexity of unidimensional functions.

Lemma 2 ([7]) Let $g:\mathbf{R} \to \mathbf{R}$ be a continuous function. The following are equivalent:
(i) g is convex

(ii) For any a, $b \in \mathbf{R}$, $g\left(\frac{a+b}{2}\right) \leq \frac{g(a) + g(b)}{2}$

(iii) For any real numbers a<b, $g\left(\frac{a+b}{2}\right) \leq \frac{1}{b-a}\int_a^b g(t)dt$.

Lemma 3 ([7]) Let $g:\mathbf{R} \to \mathbf{R}$ be a continuous function. The following are equivalent:
(i) g is convex
(ii) $g(M(X)) \leq M(g(X))$ for any random variable X.

Lemma 4 ([7]) Let $h:\mathbf{R} \to \mathbf{R}$ be integrable and $g:\mathbf{R} \to \mathbf{R}$ convex. Then

$$g(\int_a^b h(x)f(x)dx) \leq \int_a^b g(h(x))f(x)dx$$

We consider a bidimensional utility function $u:\mathbf{R}^2 \to \mathbf{R}$ of class C^2. We say that the function u(y, x) is convex in y if for any $x \in \mathbf{R}$, the unidimensional function u(., x) is convex; u(y, x) is convex in x if for any $y \in \mathbf{R}$, the unidimensional function u(y, .) is convex. Analogously one defines what means that u(y, x) is concave in y, resp. x.

Proposition 4 The following are equivalent:
(i) The function u(y, x) is convex in each of the variables y and x;
(ii) For any mixed vector (A, X), it holds:
$u(E(f, A), M(X)) \leq E_2(f, u(A, X))$
(iii) For any mixed vector (A, X), it holds:
$u(E(f, A), M(X)) \leq E_3(f, u(A, X))$.

Proposition 5 The following assertions are equivalent:
(i) The function u(y, x) is concave in each of the variables y and x.

(ii) For any mixed vector (A, X) the following inequality holds:
$u(E(f, A), M(X)) \geq E_2(f, u(A, X))$.

(iii) For any mixed vector (A, X) the following inequality holds:
$u(E(f, A), M(X)) \geq E_3(f, u(A, X))$.

Proposition 6 The following assertions are equivalent:

(i) The function u(y, x) is convex in y;

(ii) For any mixed vector (A, X),
$M(u(E(f,A),X)) \leq E_2(f,u(A,X))$;

(iii) For any mixed vector (A, X),
$M(u(E(f,A),X)) \leq E_3(f,u(A,X))$.

Proposition 7 The following assertions are equivalent:

(i) The function u(y, x) is convex in y;

(ii) For any mixed vector (A, X),
$M(u(E(f,A),X)) \geq E_2(f,u(A,X))$;

(iii) For any mixed vector (A, X),
$M(u(E(f,A),X)) \geq E_3(f,u(A,X))$.

Proposition 8 The following assertions are equivalent:

(i) The function u(y, x) is convex in x;

(ii) For any mixed vector (A, X),
$E_2(f,u(A,M(X)) \leq E_2(f,u(A,X))$;

(iii) For any mixed vector (A, X),
$E_3(f,u(A,M(X)) \leq E_3(f,u(A,X))$.

Proposition 9 The following assertions are equivalent:

(i) The function u(y, x) is concave in x;

(ii) For any mixed vector (A, X),
$E_2(f,u(A,M(X)) \geq E_2(f,u(A,X))$;

(iii) For any mixed vector (A, X),
$E_3(f,u(A,M(X)) \geq E_3(f,u(A,X))$.

IV MIXED RISK AVERSION

In this section we will study the risk aversion of an agent in front of a risk situation with two parameters: one described by a fuzzy number and another described by a random variable. In other words, the risk situation will be described by a mixed vector (A, X), in which A is a fuzzy number, and X is a random variable. The agent will be represented by a utility function u(y, x) of class C^2 with $u_1 > 0, u_2 > 0$.

A weighting function f:[0, 1]→**R** is fixed.

Proposition 10 Let $(y, x) \in \mathbf{R}^2$ and (A, X) a mixed vector with E(f, A)=0 and M(X)=0. Then for any $j \in \{1,2,3\}$ the following approximation formulas are valid:

$E_j(f,u(y+A,x+X)) \approx$

$$u(y,x)+\frac{1}{2}u_{11}(y,x)Var_j(f,A)+\frac{1}{2}u_{22}(y,x)Var(X) \qquad (11)$$

$$E_j(f,u(y+A,x)) \approx u(y,x)+\frac{1}{2}u_{11}(y,x)Var_j(f,A) \qquad (12)$$

$$M(u(y,x+X)) \approx u(y,x)+\frac{1}{2}u_{22}(y,x)Var(X) \qquad (13)$$

$E_j(f,u(y+A,x+X))-E_j(f,u(y+A,x)) \approx$

$$\frac{1}{2}u_{22}(y,x)Var_j(f,A) \qquad (14)$$

$$E_j(f,u(y+A,x+X))-M(u(y,x+X)) \approx \frac{1}{2}u_{11}(y,x)Var(X) \qquad (15)$$

In the following, for each of the three mixed expected utilities defined in Section 2, we will define the notions of risk averse, risk lover and risk neutral agent.

Formulas (12) and (13) are particular cases of (11); (14) is obtained from (11) and (12), and (15) is obtained from (11) and (13).

In any risk theory it is important to know that an agent (represented by a utility function) is risk averse, risk lover or risk neutral. In the possibilistic risk approach, these notions are defined in the framework of possibilistic EU-theory [8]. Next we assume that the agent is represented by a utility function u(y, x) of class C^2. As usual we will identify the agent with its utility function.

Lemma 5 Let $j \in \{1,2,3\}$. The following assertions are equivalent:

(i) For any $(y, x) \in \mathbf{R}^2$ and for any mixed vector (A, X) the following inequality holds:

$$E_j(f,u(y+A,x+X)) \leq u(y+E(f,A),x+M(X))$$

(ii) For any $(y, x) \in \mathbf{R}^2$ and for any mixed vector (A, X) with E(f,A)=0 and M(X)=0, one has:

$$E_f(u(y+A),x+X)) \leq u(y,x).$$

Definition 3 We say that the agent u is j-mixed risk averse if it verifies the equivalent conditions of Lemma 5.

If we consider the opposite inequality the notion of j-mixed risk lover agent is obtained, and in case of equality the notion of j-mixed risk neutral agent is obtained.

The notion of j-mixed risk aversion considers the attitude of the agent in front of the mixed risk situation (A, X). We introduce now two notions of risk aversion: one refers to the possibilistic component A, and the other refers to the probabilistic component X.

Lemma 6 Let $j \in \{1,2,3\}$. The following assertions are equivalent:

(i) For any $(y, x) \in \mathbf{R}^2$ and for any mixed vector (A, X),

$$E_j(f,u(y+A,x+X)) \leq M(u(y+E(f,A),x+X))$$

(ii) For any $(y, x) \in \mathbf{R}^2$ and for any mixed vector (A, X) with E(f, A)=0,

$$E_j(f,u(y+A,x+X)) \leq M(u(y,x+X))$$

Lemma 7 Let $j \in \{1,2,3\}$. The following assertions are equivalent:

(i) For any $(y, x) \in \mathbf{R}^2$ and for any mixed vector (A, X),

$$E_j(f,u(y+A,x+X)) \leq E_j(f,u(y+A,x+M(X)))$$

(ii) For any $(y, x) \in \mathbf{R}^2$ and for any mixed vector (A, X) with M(X)=0,

$$E_j(f, u(y+A, x+X)) \leq E_j(f, u(y+A, x))$$

Definition 4 Let $j \in \{1,2,3\}$. We say that the agent u is:

(a) j-possibilistically risk averse, if the equivalent conditions of Lemma 6 hold.

(b) j-probabilistically risk averse, if the equivalent conditions of Lemma 7 hold.

We will characterize now the three risk aversion conditions introduced by Definitions 3 and 4 by convexity conditions of u(y, x).

Proposition 11 Let $j \in \{1,2,3\}$. The following assertions are equivalent:

(i) The agent u is j-mixed risk averse;

(ii) The function u(y, x) is concave in each of the variables y and x.

Proposition 12 Let $j \in \{1,2,3\}$. The following assertions are equivalent:

(i) The agent u is j-possibilistically risk averse;

(ii) The function u is concave in y.

Proposition 13 Let $j \in \{1,2,3\}$. The following assertions are equivalent:

(i) The agent u is j-probabilistically risk averse;

(ii) The function u is concave in x.

The following result establishes the equivalence between the risk aversion of an agent in front of mixed risk and the risk aversion of the agent in front of each of its two components.

Proposition 14 Let $j \in \{1,2,3\}$. The agent u is j-mixed risk averse iff it is simultaneously j-possibilistically risk averse and j-probabilistically risk averse.

After having defined what means that an agent is mixed risk-averse it is necessary to have an indicator of the level of mixed risk aversion. In the probability theory of risk, the measurement of risk aversion is done by (probabilistic) risk premium [9], [10].

The mixed risk aversion will be evaluated by the indicators introduced by the following definition:

Definition 5 Assume that the utility function u has the class C^2 and it is strictly increasing in each argument. Let $(y, x) \in \mathbf{R}^2$ and the mixed vector (A, X), with E(f, A)=0 and M(X)=0. For any $j \in \{1,2,3\}$, we define:

• the j-possibilistic risk premium $\pi_j = \pi_j(y, x, A, X, u)$ as the unique solution of the equation:

$$E_j(f, u(y+A, x+X)) = u(y-\pi_j, x)$$

• the j-probabilistic risk premium $\rho_j = \rho_j(y, x, A, X, u)$ as the unique solution of the equation:

$$E_j(f, u(y+A, x+X)) = u(y, x-\rho_j)$$

•the j-mixed risk premium vector (α_j, β_j) as a solution of the equation:

$$E_j(f, u(y+A, x+X)) = u(y-\alpha_j, x-\beta_j)$$

Remark 2 The equation which defines the j-mixed risk premium vector may have several solutions.

Next we establish the approximation formula of the indicators introduced by Proposition 10.

Proposition 15 Assume that the utility function u has the class C^2, $u_1 > 0, u_2 > 0$. Let $(y, x) \in \mathbf{R}^2$ and the mixed vector (A, X), with E(f, A)=0 and M(X)=0. For any $j \in \{1,2,3\}$, we have:

(a) the possibilistic risk premium π_j can be approximated by the formula:

$$\pi_j \approx -\frac{1}{2} \frac{u_{11}(y,x)Var_j(f,A) + u_{22}(y,x)Var(X)}{u_1(y,x)} \tag{16}$$

(b) the probabilistic risk premium ρ_j can be approximated by the formula:

$$\rho_j \approx -\frac{1}{2} \frac{u_{11}(y,x)Var_j(f,A) + u_{22}(y,x)Var(X)}{u_2(y,x)} \tag{17}$$

(c) the mixed risk premium vector (α_j, β_j) can be approximated by the formula:

$$\alpha_j \approx -\frac{1}{4} \frac{u_{11}(y,x)Var_j(f,A) + u_{22}(y,x)Var(X)}{u_1(y,x)} \tag{18}$$

$$\beta_j \approx -\frac{1}{4} \frac{u_{11}(y,x)Var_j(f,A) + u_{22}(y,x)Var(X)}{u_2(y,x)} \tag{19}$$

Proposition 16 Keeping the conditions and notations of Proposition 15, we have:

(a) $\pi_3 \approx \pi_2 - 2\pi_1$;(b) $\rho_3 \approx \rho_2 - 2\rho_1$;

(c) $(\alpha_3, \beta_3) \approx (\alpha_2, \beta_2) - 2(\alpha_1, \beta_1)$.

REFERENCES

[1] C. Carlsson, R. Fullér, "On possibilistic mean value and variance of fuzzy numbers," Fuzzy Sets and Systems, 22, 2001, pp. 315-326.

[2] C. Carlsson, R. Fullér, Possibility for Decision. Springer Verlag, 2011.

[3] R. Fullér, P. Majlender, "On weighted possibilistic mean value and variance of fuzzy numbers," Fuzzy Sets and Systems, 136, 2003, pp. 363-374.

[4] I. Georgescu, "Possibilistic risk aversion." Fuzzy Sets and Systems, 60, 2009, pp. 2608-2619.

[5] I. Georgescu, Possibility Theory and the Risk. Springer Verlag, 2012.

[6] W. G. Zhang, Y. L. Wang, "A comparative analysis of possibilistic variances and covariances of fuzzy number," Fundamenta Informaticae, 79(1-2), 2008, pp. 257-263.

[7] C. P. Niculescu, L. E. Perrson, Convex Functions and Their Applications: A Contemporary Approach. Springer, 2005.

[8] A. M. Lucia Casademunt, I. Georgescu, "Connecting Possibilistic Prudence and Optimal Saving," International Journal of Artificial Intelligence and Interactive Multimedia, 2(4), 2013, pp. 38-45.

[9] K. J. Arrow, Essays in the Theory of Risk Bearing. Amsterdam: North Holland, 1970.

[10] J. Pratt, "Risk aversion in the small and in the large," Econometrica, 32, 1970, pp. 122-130.

Fuzzy inference system for real option valuation with the fuzzy pay-off method

Mariia Kozlova, Pasi Luukka and Mikael Collan
School of Business and Management
Lappeenranta University of Technology
Lappeenranta, Finland
mariia.kozlova@lut.fi

Abstract—**This extended abstract presents an extension of the fuzzy inference system framework for investment analysis with the fuzzy pay-off method. We build an add-on module that performs real option valuation. The new framework is able to reflect peculiarities of complex investment cases with complicated causal relationship of the influencing factors. Incorporating the fuzzy inference system and real option valuation into the investment analysis enables creation of an intelligent decision support system that equips investors with actionable information on complex investments. Basing on the simple-in-use fuzzy pay-off method, the framework can be implemented on top of existing spreadsheet investment model, facilitating its adoption throughout the business world.**

Keywords—fuzzy pay-off method; fuzzy inference system; real option valuation

I. INTRODUCTION

In this paper we introduce an advancement of the novel idea of constructing a fuzzy inference system (FIS) to complement fuzzy pay-off method-based investment analysis with a module computing a real option value of the final fuzzy net present value (NPV) distribution. This module is elaborated for the FIS framework presented in [1]. A rationale behind incorporating FIS into investment analysis is to be able to provide more actionable information with regards to the important value-affecting variables by introducing an artificial decomposition of the pay-off distribution into multiple sub-distributions, based on natural, identified, or arbitrarily set value thresholds.

Despite wide application of FIS in different engineering areas [2-7], its adoption to investment analysis is very limited. Such cases include using FIS to evaluate input parameters [8] and merging it with system dynamic modeling [9].

The framework suggests building a classical cash flow model of the investment; identifying key variables and their states; applying the fuzzy pay-off method [10] to gain NPV sub-distributions for each combination of states of the key variables; implementing FIS that bridges variable state combinations with their consecutive fuzzy NPVs and allows analyzing the overall outcome from crisp inputs of each variable. Introduction of the real option calculation into this framework complements investment analysis with real option thinking and together it

creates a comprehensive platform for analysis of complex investments.

II. THE PROPOSED ADD-ON

The FIS used in the framework is a classical Mamdani-type fuzzy inference system [11], for details see [1]. A multiple-input-single-output (MISO) fuzzy system is utilized with generalized modus ponens [12] type of inference, so that rules are formed as follows:

R_1: If $\tilde{V}_1$ is $\tilde{A}_{11}$ and $\tilde{V}_2$ is $\tilde{A}_{21}$ and ,..., and $\tilde{V}_n$ is $\tilde{A}_{n1}$ then w is NPV_1
$\vdots$
R_K: If $\tilde{V}_1$ is $\tilde{A}_{1K}$ and $\tilde{V}_2$ is $\tilde{A}_{2K}$ and,..., and $\tilde{V}_n$ is $\tilde{S}_{nK}$ then w is NPV_K
fact: $\tilde{V}_1$ is $\bar{v}_{01}$ and $\tilde{V}_2$ is $\bar{v}_{02}$ and ,..., and $\tilde{V}_n$ is $\bar{v}_{0n}$
consequence: w is NPV

where $\tilde{V}_1$, $\tilde{V}_2$,..., $\tilde{V}_n$ are the key variables, $\tilde{A}_{ik}$ is i^{th} state of the kth rule taken from pool of relevant states $\tilde{S}_{ij}$ defined for the kth group. NPV_k is the fuzzy NPV distribution for each group resulting from applying the fuzzy pay-off method for each combination of states of key variables.

The consequence is computed by

$$consequence = Agg(fact \circ R_1, ..., fact \circ R_K) \quad (1)$$

Computing fuzzy output can be formulated in three steps:

i) Computing firing level α_k of the k-th rule by
$$\alpha_k = \tilde{V}_1(\bar{v}_{01}) \wedge \tilde{V}_2(\bar{v}_{02}) \wedge, ..., \wedge \tilde{V}_n(\bar{v}_{0n}) \quad (2)$$

ii) Computing the output of the k-th rule by
$$NPV'_k(w) = \alpha_k \wedge NPV_k(w) \quad (3)$$

iii) Overall system fuzzy output NPV is computed from individual rule outputs $NPV'_k(w)$ by
$$NPV(w) = NPV'_1(w) \vee NPV'_2(w) \vee \cdots \vee NPV'_K(w) \quad (4)$$

For the intersection and union operators we use minimum and maximum.

The real option value computation of the new module consists of three steps:

i) Defining the positive part of the fuzzy output NPV
$$NPV^+(w) = max\{0, NPV(w)\} \quad (5)$$

The authors would like to acknowledge the financial support received by M. Kozlova from Fortum Foundation.

ii) Computing the success ratio that is a ratio of the positive area of the fuzzy NPV to its whole area

$$r_{success} = \frac{\int_w NPV^+(w)dw}{\int_w NPV(w)dw} \qquad (6)$$

iii) Computing real option value *ROV* as a center of gravity of the positive area weighted on the success ratio

$$ROV = \frac{\int_w wNPV^+(w)dw}{\int_w NPV^+(w)dw} r_{success} \qquad (7)$$

The ROV computation is realized as a defuzzification operator for FIS and coded in Matlab® as follows:

```
function defuzzfun =
total_area = sum(ymf); % the whole area of the fuzzy
positive_area = sum(ymf(xmf>0)); % positive area of
fuzzy
mean_positive =
% expected
success = positive_area/total_area; % success
defuzzfun = mean_positive*success; % real option
end
```

Published with MATLAB® R2015b

III. NUMERICAL ILLUSTRATION AND CONCLUSION

The application of the proposed approach is illustrated on the real-world investment case of industrial scale solar PV power plant under Russian renewable energy support policy. The details on the case can be found in [13, 14].

Following the framework [1], the inputs and outputs to the FIS are defined (Fig. 1). The influential manageable factors include capital costs (CapEx), capacity factor (CapF) or electricity production performance, and localization (Loc) that is a share of locally manufactured equipment in the project. The states of these variables are defined in accordance with the thresholds set by the supporting policy [15], which define an amount of support received by a project. The output is

constructed based on the pay-off method run for each combination of states of input variables.

The rule base simply reflects the scenario formation and the fuzzy NPV output from FPOM implementation. As the result, constructed system provides a rule viewer interface (Fig. 2), where one can see how different combinations of crisp values of the key variables feed the FIS firing outputs of different rules and resulting in an output fuzzy NPV of a complex shape. The system returns the real option value computed with the proposed add-on.

By 'playing' with the built fuzzy inference tool, decision-makers could estimate acceptable uncertainty levels for each key variable and elaborate an actionable plan to achieve desired profitability of investment. Constructed FIS delivers more graphical analytics omitted in this extended abstract.

To conclude, the presented module complements the FIS framework for investment analysis with the fuzzy pay-off method enabling real option computation. Apart from representing the holistic picture of an investment, FIS can serve as a tool for a particular project analysis with defined prospective values of uncertain variables. If the latter represent project parameters that can be to some extent influenced by project managers, the framework becomes a roadmap that navigates investors in search of project profitability.

REFERENCES

[1] M. Kozlova, P. Luukka and M. Collan, "A new method for deeper analysis of investments: Application of a fuzzy inference system into investment analysis and decision-making," (unpublished).

[2] G. Feng, "A survey on analysis and design of model-based fuzzy control systems," Fuzzy Systems, IEEE Transactions On, vol. 14, pp. 676-697, 2006.

[3] M. Fazzolari, R. Alcala, Y. Nojima, H. Ishibuchi and F. Herrera, "A review of the application of multiobjective evolutionary fuzzy systems: Current status and further directions," Fuzzy Systems, IEEE Transactions On, vol. 21, pp.

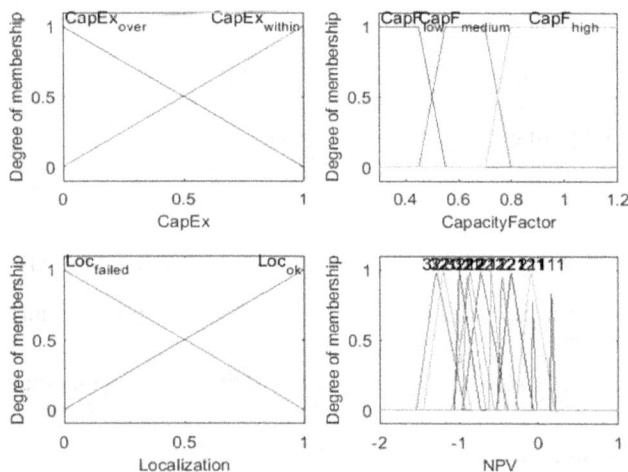

Fig. 1. Inputs and outputs to FIS

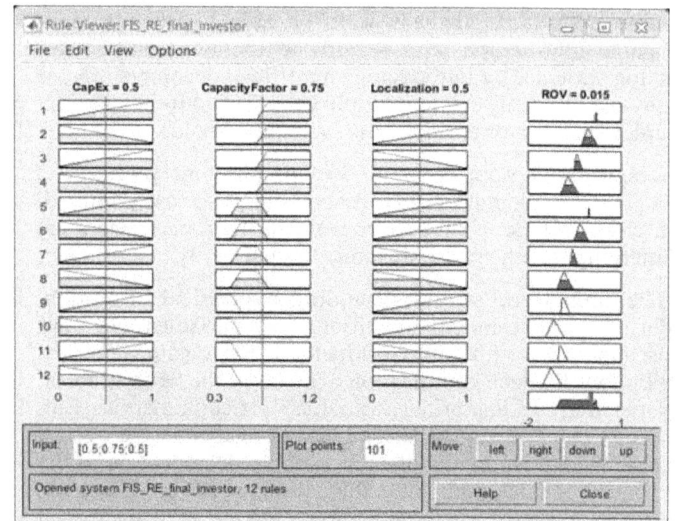

Fig. 2. Fuzzy inference system visualization

[5] C. Chiou, C. Chiou, C. Chu and S. Lin, "The application of fuzzy control on energy saving for multi-unit room air-conditioners," Appl.Therm.Eng., vol. 29, pp. 310-316, 2009.

[6] J. CUI, J. TAN, L. AO and W. KANG, "Automatic focusing method based on fuzzy control and image processing technique [J]," Opto-Electronic Engineering, vol. 6, pp. 004, 2005.

[7] T. Das and I.N. Kar, "Design and implementation of an adaptive fuzzy logic-based controller for wheeled mobile robots," Control Systems Technology, IEEE Transactions On, vol. 14, pp. 501-510, 2006.

[8] A. Ustundag, M.S. Kılınç and E. Cevikcan, "Fuzzy rule-based system for the economic analysis of RFID investments," Expert Syst.Appl., vol. 37, pp. 5300-5306, 2010.

[9] A. Arasteh and A. Aliahmadi, "A proposed real options method for assessing investments," The International Journal of Advanced Manufacturing Technology, vol. 70, pp. 1377-1393, 2014.

[10] M. Collan, R. Fullér and J. Mezei, "A fuzzy pay-off method for real option valuation," Journal of Applied Mathematics and Decision Sciences, vol. 2009, pp. 1-14, 2009.

[11] E.H. Mamdani and S. Assilian, "An experiment in linguistic synthesis with a fuzzy logic controller," International Journal of Man-Machine Studies, vol. 7, pp. 1-13, 1975.

[12] L.A. Zadeh, "Outline of a new approach to the analysis of complex systems and decision processes," Systems, Man and Cybernetics, IEEE Transactions On, pp. 28-44, 1973.

[13] M. Kozlova, "Analyzing the effects of the new renewable energy policy in Russia on investments into wind, solar and small hydro power," Master's thesis, Lappeenranta University of Technology, pp. 104, 2015.

[14] M. Kozlova, M. Collan and P. Luukka, "Comparing Datar-Mathews and fuzzy pay-off approaches to real option valuation," in *Real Option Workshop*, pp. 29-34, 2015.

[15] Government of Russian Federation, "28 May 2013 Decree #449 on the mechanism of promoting the use of renewable energy in the wholesale market of electric energy and power," 2013.

Likert scales in group multiple-criteria evaluation: a distance-from-ideal based approach

Jan Stoklasa
and Tomáš Talášek
Lappeenranta University of Technology
School of Business and Management
and Palacký University Olomouc
jan.stoklasa@lut.fi
tomas.talasek@lut.fi

Jaroslava Kubátová
and Klára Seitlová
Palacký University Olomouc
Faculty of Arts
Department of Applied Economics
jaroslava.kubatova@upol.cz
klara.seitlova@upol.cz

Abstract—Likert scales are a widely used tool for the evaluation and attitude expression in many fields of social science. In this paper we explore their use in multiple-criteria multi-expert evaluation. We propose a methodology that deals with the non-uniformity of the distribution of linguistic labels along the evaluation universe and also with possible response bias (central tendency and extreme-response tendency). The methodology represents the Likert-type evaluations of an alternative with respect to various criteria using histograms. Histograms are used to present information and also in the process of aggregation of information, since the underlying evaluation scale is ordinal. A transformation of the multi-expert multiple-criteria evaluation represented by a histogram into a 3-bin histogram to control for the response bias is performed and an ideal-evaluation 3-bin histogram is defined with respect to the number of criteria and number of evaluators. We propose a distance measure to assess the closeness of the overall evaluation to the ideal. The value of the distance measure can be effectively used in fuzzy-rule-based systems to provide an interpretation extension to the Likert-type evaluation tools. We discuss the possible uses in management and marketing research and in psychological diagnostics.

Keywords—Likert scale, histogram, distance, evaluation, MCDM, group evaluation.

I. INTRODUCTION

Likert scales were introduced by Likert in 1930s [1] as a tool for the measurement and assessment of attitudes. Since then Likert scales (and Likert-type scales) have grown popular in many fields of social science, including management and marketing research [2]. They became a frequently used tool for the extraction of information from participants concerning not only their attitudes, but also preferences, evaluations etc.

There seems to be enough reasons for not considering the values provided by Likert scale items in the questionnaire as interval or ratio scale values in the process of their aggregation. The meaning of the linguistic labels can be context dependent (hence the distances between them expressed in the numerical level can differ). Aggregation of information provided by the Likert scale can therefore be problematic (in the sense of the necessary restriction to ordinal computational methods). Also the well known response biases in self-report questionnaires, mainly the extremity response (*leniency*) tendency and the midpoint response (*central*) tendency of the decision makers (see e.g. [2]) can complicate the aggregation.

II. LIKERT SCALES

The *n-point Likert scale* [1] can be defined as a discrete bipolar evaluation scale with n integer values (e.g. $[1, 2, 3, 4, \ldots, n]$). The values are naturally ordered, the endpoints 1 and n are interpreted as extreme evaluations (1 as completely positive, n as completely negative, or vice versa). The Likert scale can have two interlinked levels of description - the numerical one represented by the integer values and a linguistic level assigning each (or some) of the values a linguistic label (see Fig. 1). In practical applications either both levels, or just one of the levels is provided to the decision makers.

Strong positive	Positive	Neutral	Negative	Strong negative
(1)	(2)	(3)	(4)	(5)

Strongly agree	Agree	More or less agree	Undecided	More or less disagree	Disagree	Strongly disagree
(1)	(2)	(3)	(4)	(5)	(6)	(7)

Fig. 1. An example of a 5-point and 7-point Likert Scale.

III. PROBLEM DEFINITION AND SUGGESTED SOLUTION

In this paper we discuss Likert scales (see Fig. 1) in the context of multiple-criteria and multi-expert evaluation. As such, also the area of psychology, management, marketing and economical research, behavioral science, sociology and related fields are the possible recipients of the presented results. We suggest a way to aggregate individual evaluations (provided by Likert scales) into an overall evaluation and a way of aggregating these overall evaluations across experts that reflect the specifics of Likert scales.

After an inspection of the available histogram distances we conclude that neither the nominal-histogram distances nor the ordinal-histogram distances are directly applicable in the Likert scale setting if the above mentioned issues of possible response bias and the non-uniformity of the scale is considered. On reduced 3-bin histograms we introduce the "ideal evaluation", which can be interpreted in the linguistic form as "all the answers are close to the desired pole of the scale" (e.g. "all the answers are 1 or 2 in a 7-point Likert scale, where 1 is the best evaluation"). Subsequently we calculate the distance of the overall group evaluation represented by a 3-bin histogram to this ideal taking into account ordinality, response bias and the symmetry of Likert scales with respect to their middle point. Our approach results in an absolute-type evaluation (see e.g. [3], [4]) with the interpretation of a degree of fulfillment of a given goal. This value can then be easily used in fuzzy rule based systems [5], that can provide interpretation, diagnostics and summarizing extensions to the outputs of the evaluation model.

IV. Conclusion

The main contribution of this paper is the suggestion of a methodology of using Likert scales that respects the ordinal character of these scales (the elements of the Likert scale may not be equidistant, but they need to be ordered). Using histograms for the representation of aggregated values of Likert scales (overall evaluation for 1 individual, overall evaluation across all individuals involved), we take the differences in decision-makers' responses into account (extremity responses, mid-point responses) by joining specific bins of the histograms.

We build on the symmetry of Likert scale with respect to the middle of the continuum on which it is defined and introduce a distance among thus constructed histograms. The definition of an ideal evaluation also reflects the response-bias. We suggest a tool that not only respects the specific features of Likert scales using numerical and linguistic labels of their elements, but also provides outputs that can be used in further analysis. These outputs can be interpreted as degrees of fulfillment of a given goal (it is an example of an evaluation of absolute type [3], [4]) and can be further used in fuzzy-rule-based systems. This is an important feature of the proposed multiple-criteria multi-expert use of Likert scale for evaluation purposes, since it can provide easy to design and easy to use tools for the automation of interpretation of the information obtained through Likert scales.

V. Acknowledgments

This research was partially supported by the grant IGA_FF_2016_007 Continuities and Discontinuities of Economy and Management in the Past and Present 2.

References

[1] R. Likert, "A technique for the measurement of attitudes," *Archives of Psychology*, vol. 22, no. 140, pp. 1–55, 1932.

[2] G. Albaum, "The Likert scale revisited : An alternate version," *Journal of the Market Research Society*, vol. 39, no. 2, pp. 331–348, 1997.

[3] V. Jandová and J. Talašová, "Evaluation of absolute type in the Partial Goals Method," in *Proceedings of the 33rd International Conference on Mathematical Methods in Economics*, pp. 321–326, 2015.

[4] J. Talašová, "NEFRIT-Multicriteria decision making based on fuzzy approach.," *Central European Journal of Operations Research*, vol. 8, no. 4, pp. 297–319, 2000.

[5] J. Stoklasa, T. Talášek, and J. Musilová, "Fuzzy approach - a new chapter in the methodology of psychology?," *Human Affairs*, vol. 24, pp. 189–203, mar 2014.

Estimating Capital Requirements for One-Off Operational Risk Events

Extended Abstract

Yuri Lawryshyn
Centre for Management of Technology and
Entrepreneurship (CMTE)
University of Toronto
Toronto, Ontario, Canada
yuri.lawryshyn@utoronto.ca*

Pasi Luukka
School of Business and Management
Lappeenranta University of Technology
Lappeenranta, Finland

Fai Tam
CMTE
University of Toronto
Toronto, Ontario, Canada

Jerry Fan
CMTE
University of Toronto
Toronto, Ontario, Canada

Mikael Collan
School of Business and Management
Lappeenranta University of Technology
Lappeenranta, Finland

Abstract— **Under current regulations, set by Basel II, banks are required to possess enough capital for a 99.9% annual Value-at-Risk (VaR) event for operational risk. Currently, our client, "the Bank", uses two methods to calculate their 99.9% VaR, as part of the Basel II guideline: through a statistical approach, and a scenario analysis approach. Our objective is to create a framework based on both a statistical approach and a scenario analysis to estimate the expected loss from severe and infrequent events from an operational risk perspective.**

Keywords—One off events; operational risk; value at risk; fuzzy logic, extreme value theorem

I. INTRODUCTION

Under current regulations, set by Basel II, banks are required to possess enough capital for a 99.9% annual Value-at-Risk (VaR) event for operational risk. This value could also be loosely interpreted as the unexpected loss associated with a once-in-a-millennium operational loss event. Currently, our client, "the Bank", uses two methods to calculate their 99.9% VaR, as part of the Basel II guideline: through a statistical approach, and a scenario analysis approach. In the statistical approach, losses are assumed to follow a statistical distribution, and the VaR is calculated based on the distribution tail. In the scenario analysis approach, subject-matter experts (SMEs) review historical operational risk incidents and corroborate on potential future losses. Our objective is to create a framework based on both a statistical approach and a scenario analysis to estimate the expected loss from severe and infrequent events from an operational risk perspective.

II. METHODOLOGY

Lu [1] determined the capital requirement due to operational risk, in compliance with Basel II, of Chinese commercial banks using a bottom-up approach. He gathered publicly available pooled data on the frequency of incidences as well as gross loss by business line and by event type for all the Chinese banks. One issue encountered by Lu, as well as the Bank, is the scarcity of data. Lu resolved the problem by bootstrapping the data. Using the loss data, Lu was able to calculate the tail risk by fitting the generalized Pareto distribution (GPD) through the use of the peak-over-threshold. Bensalah [2] purports the use of the GPD for estimating extreme out of sample cases. Our approach will be to utilize the GPD as well as the log-normal distribution and bootstrapping to estimate the 99.9% tail VaR.

Fuzzy set theory provides a natural methodology to utilize linguistic information to estimate numerical outcomes. It seems obvious that the use of fuzzy logic can be used in scenario analysis. According to Reveiz and Leon [3], while there are several studies that support the use of fuzzy logic in operational risk management and create theoretical models for it, there is very little literature on the application of the fuzzy models in operational risk. One such example is the fuzzy logic model proposed by Cerchiello and Giudi (2013) by quantifying IT operational risk for a telecom company in Israel.

TABLE I.   99.9% ANNUAL VaR AND ITS 95% CONFIDENCE INTERVAL

Distribution	99.9% Annual VaR	95% Confidence Interval
Lognormal	$88.18 million	[$12.36 million, $1.55 billion]
GPD	$432.2 million	[$1.79 million, $10.2 quadrillion ($10^{16}$)]

TABLE II.   PREDICTED 9-YEAR WORST LOSS. THE ACTUAL WORST LOSS IS 6,678,000.

Distribution	Predicted 9 Year Worst Loss	95% Confidence
Lognormal	$3.55 million	[957k, 21.1 million]
GPD	$13.25 million	[1.7294 million, 12.4 billion]

In this study, we utilize the lognormal and GPD distributions to fit historical data provided to us by the Bank. It should be noted that the historical operational loss data consisted of only 29 points over a span of nine years. Thus, estimating the distribution parameters is challenging. We followed the bootstrapping approach of Lu [1] to deal with the lack of data.

Given the sparseness of available data, we analyzed 100 years of flood data to test our approach. A rolling window back test with 1 year of window data to predict the 50-year worst case was performed. Finally, to demonstrate the sensitivity of the confidence interval relative to the number of data points, we used a cumulative window to predict the 10-year worst case. The flood data analysis provides a proxy for the accuracy of implementing Lu's [1] methodology for the sparse data that is available.

Our fuzzy logic model is based on the framework of Radionovs et al (2014). We developed a fuzzy logic framework for the Bank to use for their scenario analysis.

III. RESULTS

By fitting the Bank loss data, we calculated the Bank's 99.9% VaR for operational risk and estimated the 95% confidence interval of the VaR value. Table 1 summarizes the results. As can be seen, the results are highly variable. To test the validity of our approach, we removed the 9-year largest loss and tested the ability of our proposed approach to estimate this

loss of $6.69 million (see Table 2). As can be seen, the lognormal distribution underestimated the loss while the GPD overestimated it.

As mentioned above, we used flood data to better understand the validity of our approach. The data consisted of 34,086 observations of discharge data of the Ramapo River (New York State), recorded between 1922 and 2015. Discharge rate was used to measure the likelihood of a flood as flooding occurs at high discharge rates. First, we used one year of data to predict the 50-year highest discharge through a rolling window analysis. Figure 1 compares the predicted and actual worst cases using the different distributions. As can be seen the lognormal distribution is the better predictor of extreme events compared to the GPD. In addition, the GPD has such extreme confidence intervals that its predictions are suspect.

Next, we used a cumulative analysis, where, as more data is made available, all data is used in the prediction (Figure 2). In this case, we see that the GPD not only is the better estimator of extreme event, but also the predicted worst case is always larger than the actual extreme event. This analysis suggests that the GPD can be used as an upper bound of extreme loss. Furthermore, the confidence interval reduces substantially as the number of data points increases. As a result, we believe the GPD performs better than the log normal distribution when there is a large set of data.

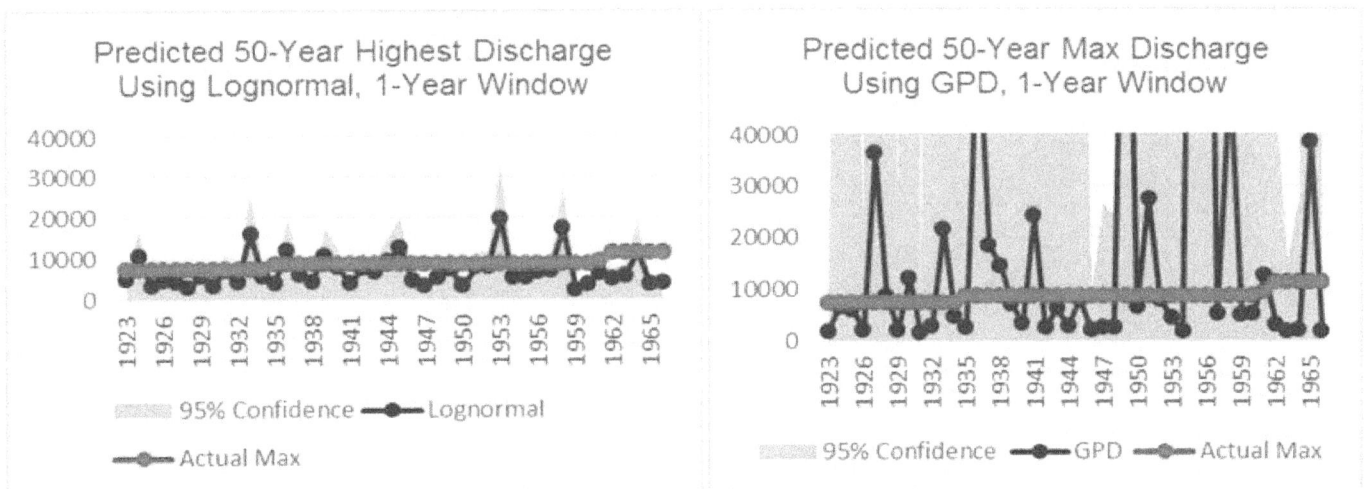

Fig. 1.   Prediction of 50 year worst case event using 1-year data.

Proceedings of NSAIS16 - 2016 Lappeenranta Finland - ISBN 978-952-265-986-6

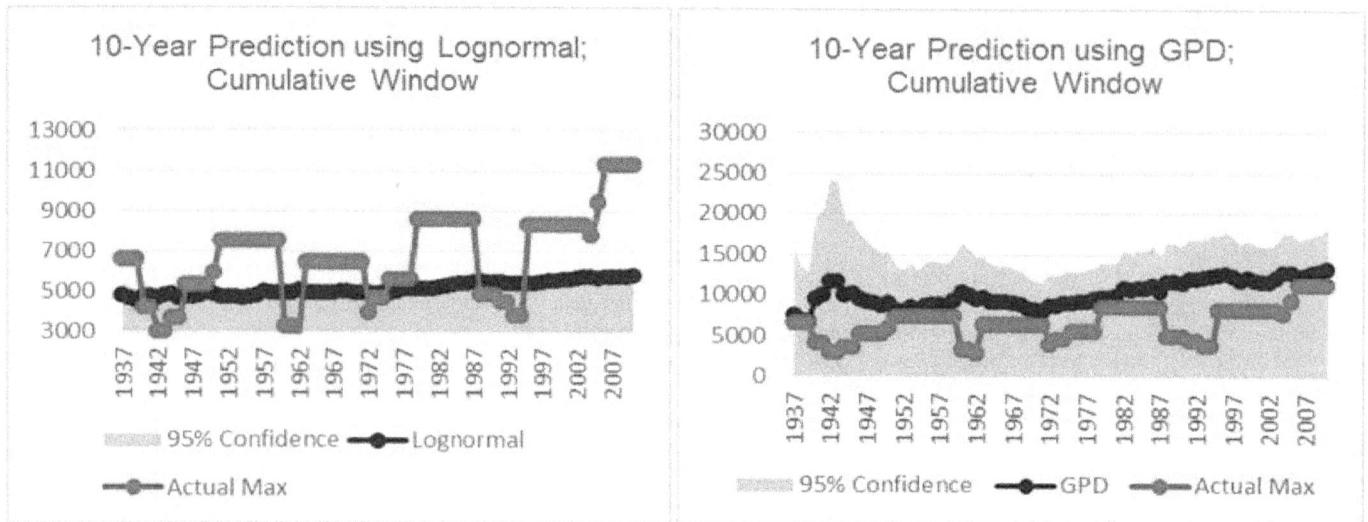

Fig. 2. Cumulative analysis of the flood / discharge data

The Bank presented us with data from their most recent SME scenario analysis. Using this data, we developed a fuzzy logic framework. The framework consists of using low, medium and high loss estimates provided by the SMEs. In the framework, the SMEs are asked to rate their confidence in the estimates they provide. Furthermore, the SMEs are rated based on their banking experience. Accounting for both the SME confidence and expiring and utilizing a consensus analysis, we calculate a single fuzzy number for each scenario.

IV. SUMMARY

In this summary, we have provided two models for the Bank to use to estimate extreme operational losses. Our recommendation is to utilize the lognormal distribution for the calculation of a suitable VaR value, until more data is made available. We recommend the GPD be used to estimate upper bound of such a loss. We estimate that in a number of years (or if more data is made available) the GPD should replace the lognormal. Our fuzzy framework is unique and easy to implement. It provides SMEs the opportunity to provide low-medium-high estimates as opposed to a single value. Furthermore, their confidence in their estimate is accounted for, providing a more robust estimate.

REFERENCES

[1] Z. Lu, "Modeling the yearly value-at-risk for operational risk in Chinese commercial banks", Mathematics and Computers in Simulation, vol. 82(4), pp. 604-616, December, 2011.

[2] Y. Bensalah, "Steps in applying extreme value theory to finance: a review", Bank of Canada Working Paper 2000-20, November, 2000.

[3] A. Reveiz, and C. León, "Operational risk management using a fuzzy logic inference system", Minimise risk, optimise success working paper, September, 2009.

[4] P. Cerchiello, and P. Giudici, "Fuzzy methods for variable selection in operational risk management", The Journal of Operational Risk, vol. 7(4), pp. 25-41, December, 2012.

Approximate operations on fuzzy numbers: are they reliable?

Matteo Brunelli
Systems Analysis Laboratory
Aalto University, Finland
Email: matteo.brunelli@aalto.fi

József Mezei
Faculty of Social Sciences, Business and Economics
Åbo Akademi, Finland
Email: jmezei@abo.fi

I. Introduction

Fuzzy numbers have been extensively used to represent uncertain parameters in mathematical modelling in, for instance, decision theory [3] and operations research [10]. They can be seen as generalizations of the concept of real numbers, and therefore one natural question regards the extension of basic arithmetical operations such that these latter can operate on fuzzy numbers too. Such an extension would allow for addition, subtraction, product and division of fuzzy numbers and represent a milestone in the development of the so-called fuzzy arithmetics. It has been established in the literature that the standard extension should follow Zadeh's extension principle [9].

Specifically, considering two fuzzy numbers A_1 and A_2, the four arithmetic operations in two variables are defined as

$$(A_1 + A_2)(y) = \sup_{y = x_1 + x_2} \min\{A_1(x_1), A_2(x_2)\} \quad (1)$$

$$(A_1 - A_2)(y) = \sup_{y = x_1 - x_2} \min\{A_1(x_1), A_2(x_2)\} \quad (2)$$

$$(A_1 \cdot A_2)(y) = \sup_{y = x_1 \cdot x_2} \min\{A_1(x_1), A_2(x_2)\} \quad (3)$$

$$(A_1/A_2)(y) = \sup_{y = x_1/x_2} \min\{A_1(x_1), A_2(x_2)\} \quad (4)$$

Note that the operations (1)–(4) are compatible with the possibilistic interpretation of fuzzy sets [4], meaning that (1)–(4) are mathematically sound also when the fuzzy numbers are used as representations of possibility distributions. Additionally, extension principle-based operations are also compatible with the α-cut representation (and the decomposition theorem) of fuzzy numbers [7].

Considering that the most used shapes of fuzzy numbers are triangular and trapezoidal (piecewise linear functions), it is extremely relevant to note that the operations of product (3) and division (4) of triangular and trapezoidal fuzzy numbers do *not* preserve linearity. Nonetheless, neglecting this result, countless contributions to the literature employ approximate and linear preserving multiplication and division for fuzzy numbers. In the following a triangular fuzzy number is represented as a triple (a, b, c), with b as the center of the fuzzy number and a and c as the left and right end-points of the support, respectively. The approximate multiplication and division of two triangular fuzzy numbers $A = (a_1, a_2, a_3)$ and $A' = (a'_1, a'_2, a'_3)$ with $a_1, a'_1 > 0$ can be defined as:

$$A \cdot A' = (a_1 \cdot a'_1, a_2 \cdot a'_2, a_3 \cdot a'_3) \quad (5)$$

$$A/A' = (a_1/a'_3, a_2/a'_2, a_3/a'_1) \quad (6)$$

Said this, it is remarkable that in the literature one cannot find any study on the reliability of approximate operations on fuzzy numbers beside the study by Giacchetti and Young [5]. In the article, the authors acknowledge the problematic nature of using approximations, but they focus on developing a new approximation and do not attempt to offer a comprehensive evaluation of the magnitude of error beside some example cases. Hence, questions such as "how wrong the results can be when we use approximate operations instead of exact ones?" or "are approximate operations reliable?" are nowadays still unanswered. The lack of similar investigations is the motivation for our study, in which we try to assess the reliability of approximate operations compared to exact ones.

II. Methodology

In our research we shall adopt a pragmatic approach and consider the numerous applications of fuzzy numbers in decision making where the numbers capture subjective uncertainty. In these processes a fundamental step is that of ranking fuzzy numbers from the greatest to the smallest, i.e. establish an order relation. By means of numerical simulations we shall see how different the orders on a set of fuzzy numbers are, when one order is on exact products and the other one is on approximate products of the same set of fuzzy numbers. Equal orders would mean that, in practical terms, approximate operations are not too distortive. Conversely, different orders would indicate that when using approximate operations we are more likely to make mistakes in decision processes. In the study we considered four ranking methods:

- Center of gravity introduced in [11],

$$\mathsf{COG}(A) = \int x A(x) \mathrm{d}x \Big/ \int A(x) \mathrm{d}x . \quad (7)$$

- Possibilistic mean, originally defined by Goetschel and Voxman [6] and later studied by Carlsson and Fullér [2],

$$\mathsf{E}(A) = \int_0^1 \alpha (A_\alpha^+ + A_\alpha^-) \mathrm{d}\alpha . \quad (8)$$

- Adamo's approach [1]

$$\mathsf{AD}_\alpha(A) = A_\alpha^+ , \quad (9)$$

where A_α^+ is the rightmost point of the α-cut of A
- The method introduced by Yager in [8]

$$\mathsf{F}(X) = \frac{1}{2} \int_0^1 (A_\alpha^+ + A_\alpha^-) \mathrm{d}\alpha . \quad (10)$$

The original basic experiment was performed on a dataset of products of two randomly sampled triangular fuzzy numbers with support in the interval $[0, 1]$. However, the study is extended to the case of multiple factors and the relation between number of factors and distortion is analyzed. Furthermore, the same experiment was then repeated for the case of quotients of two fuzzy numbers. Figure 1 shows some graphical results for the case of the product. Specifically, for ranking the products the possibilistic mean [6], [2] of each product was calculated and then used to establish the order relation on them.

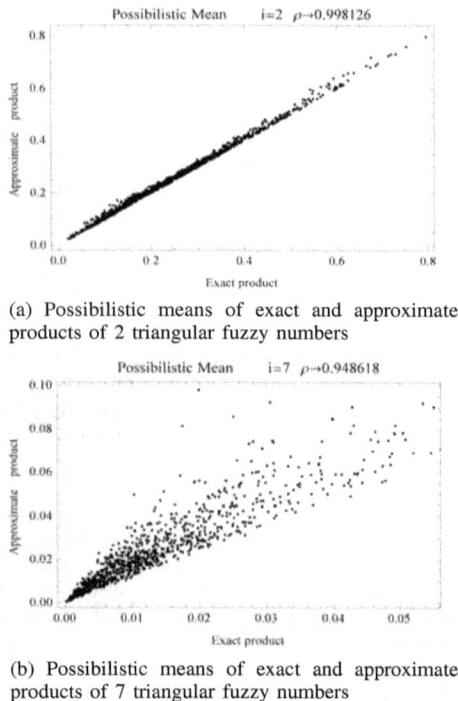

(a) Possibilistic means of exact and approximate products of 2 triangular fuzzy numbers

(b) Possibilistic means of exact and approximate products of 7 triangular fuzzy numbers

Fig. 1. The effect of approximate multiplication of an increasing number of triangular fuzzy numbers. Each point in a plot represents the possibilistic mean of the product of i triangular fuzzy numbers. The x and y axes refers to the values of the possibilistic mean when exact and approximate multiplications are used, respectively.

We will be able to show that the extent of the distortion produced by the appoximate operations is in effect connected to many factors. If for the product with two factors the approximation error is very small, this latter usually grows together with the number of factors. The extent of the error also depends on the method used to obtain the ranking of fuzzy numbers. Our work wishes to clarify all these interrelations by means of a numerical analysis and graphical results. In particular we will show that the errors in the case of approximate division can be much greater than those in the case of approximate multiplication. Thus, if on one hand our study shows that errors are small in the case of multiplication with two factors, on the other it points out that using approximate division can be extremely dangerous.

III. RESULTS

Based on the results obtained in our numerical study, we attempt to identify situations when (i) the use of approximate

operations can be an acceptable choice (although it is never advisable), and (ii) the use of approximate operations yields dangerously misleading results. Even though the decision on whether to use approximate operations ultimately depends on the error the analyst is willing to bear, we can propose some guidelines based on our statistical analysis.

G1: In the case of multiplication with two factors—which is by far the most common case in applications—the approximate operations yield results similar to the exact ones. For the four ranking methods used in this study, the Spearman index has always been greater than 0.99.

G2: If the center of gravity is employed to rank the products, then the approximate operations remain reliable, even for an increasing number of factors, with Spearman index greater than 0.998. This stability does not hold for the other three considered ranking methods.

G3: We found out that the computational time necessary to perform exact operations is actually very small. Therefore, unless results are required in a fraction of second, computational complexity is not a valid justification for using approximate operations.

G4: Except for very few choices of the support of the fuzzy numbers, the approximate division between triangular fuzzy numbers produces severe outliers and thus its use is deprecable.

G5: As the defuzzified value of approximate products tend to overestimate the defuzzified value of exact ones, when the value itself (and not only the ranking) is used as the basis of a solution to a problem, exact product should be used.

Considering that the distortion is on average greater for triangular fuzzy numbers than for trapezoidal ones, guidelines **G1**–**G3** also apply to the case of trapezoidal fuzzy numbers. Nevertheless, if trapezoidal fuzzy numbers are employed, for the same reasons, the negative effects presented in **G4** and **G5** are mitigated in proportion to the widths of their cores.

REFERENCES

[1] Adamo J.M., Fuzzy decision trees, *Fuzzy Sets and Systems*, 4(3), 207–219, 1980.

[2] Carlsson, C. and Fullér, R., On possibilitic mean value and variance of fuzzy numbers, *Fuzzy Sets and Systems*, 122(2), 315–326, 2001.

[3] Dubois D., The role of fuzzy sets in decision sciences: Old techniques and new directions. *Fuzzy Sets and Systems*, 184(1), 3–28, 2011.

[4] Dubois D. and Prade H., *Possibility Theory: An Approach to Computerized Processing of Uncertainty*, Plenum Press, 1988.

[5] Giacchetti R.E. and Young R.E., Analysis of the error in the standard approximation used for multiplication of triangular and trapezoidal fuzzy numbers and the development of a new approximation, *Fuzzy Sets and Systems*, 91(1), 1–13, 1997.

[6] Goetschel, R. and Voxman, W., Elementary fuzzy calculus, *Fuzzy Sets and Systems*, 18(1), 31–43, 1986

[7] Klir, G. and Yuan, B., *Fuzzy Sets and Fuzzy Logic: Theory and Applications*, New Jersey: Prentice Hall (1995)

[8] Yager R.R., A procedure for ordering fuzzy subsets of the unit interval, *Information Sciences*, 24(2), 143–161, 1981.

[9] Zadeh L.A., The concept of a linguistic variable and its application to approximate reasoning - I, *Information Sciences*, 8(3), 199–249, 1975.

[10] Zimmermann, H.-J., Using fuzzy sets in operational research, *European Journal of Operational Research*, 13(3), 201-216, 1983.

[11] Østergaard, J.J., *Fuzzy logic control of a heat exchanger process*, Stærkstrømsafdelingen, Danmarks Tekniske Højskole (1976).

Connection between triangular or trapezoidal fuzzy numbers and ideals

Paavo Kukkurainen

Lappeenranta University of Technology

Email: paavo.kukkurainen@lut.fi

Abstract—Abstract: By means of triangular and trapedzoidal fuzzy numbers we will construct a set $\overline{F}$ of a certain kind fuzzy numbers. Also we form a set $\overline{G}$ a certain kind of ideals in this special case. It is done in Section 4. Then $\overline{F}$ and $\overline{G}$ can be equipped with operations union $\cup$ and intersection $\cap$ (for fuzzy numbers), and with join $\vee$ and meet $\wedge$ (for ideals), respectively. Then we will introduce a lattice isomorphism between $(\overline{F}, \cup, \cap)$ and $(\overline{G}, \vee, \wedge)$. In the construction it is used some results of B. Belluce and A. Di Nola concerning Łukasiewitz rings starting with commutative rings. We need basic algebraic concepts and preliminaries of fuzzy set theory.

I. Introduction

L.P. Belluce and A. Di Nola considered Lukasiewicz rings in [1] . A starting point was commutative rings whose ideals satisfy a certain condition. We will show that this condition holds for a quotient ring $\mathbb{R}/(a)$ where (a) is a principal ideal generated by a. $\mathbb{R}$ are real numbers. We apply some results of Lukasiewicz rings to $\mathbb{R}/(a)$ which is a field and its ideal can be thought to be represented by polynomials of the first degree . On the other hand, triangular and trapedzoidal fuzzy numbers can be constructed by lines. So, we can consider fuzzy numbers which are constituted by triangular and trapedzoidal fuzzy numbers and create a connection between them and a certain kind of ideals. In fact, it is possible to set $R_{\mathcal{A}} = \mathbb{R}/(a_1) \times \mathbb{R}/(a_2) \times \cdots \times \mathbb{R}/(a_n)$ and so $Id(R_{\mathcal{A}}) = Id(\mathbb{R}/(a_1)) \times Id(\mathbb{R}/(a_2)) \times \cdots \times Id(\mathbb{R}/(a_n))$ where $Id(R_{\mathcal{A}})$ and $Id(\mathbb{R}/(a_i))$ are ideals of Lukasiewicz rings $R_{\mathcal{A}}$ and $\mathbb{R}/(a_i)$. According to the interpretation of a fuzzy set A in Section 3 we can assume that it is of the form $\mathcal{A} = (a_1, \cdots, a_n)$. The main result is represented in Proposition IV.4.

II. Preliminaries

A. Basic algebraic concepts

For basic concepts we refer to [2]

Definition II.1. A ring is a set R equipped with binary operations + and · satisfying the following three axioms, called the ring axioms:

1) $(R, +)$ is an Abelian group

- $(a + b) + c = a + (b + c)$ for all $a, b, c \in R$
- $a + b = b + a$
- There is an element $0 \in R$ such that $a + 0 = a$ for all $a \in R$
- For each $a \in R$ there exists $-a \in R$ such that $a + (-a) = 0$.

2) $(R, \cdot)$ is a monoid

- $(a \cdot b) \cdot c = a \cdot (b \cdot c)$ for all $a, b, c \in R$
- There is an element $1 \in R$ such that $a \cdot 1 = a$ and $1 \cdot a = a$ for all $a \in R$

3) Multiplication is distributive with respect to addition

- $a \cdot (b + c) = (a \cdot b) + (a \cdot c)$ for all $a, b, c \in R$
- $(b + c) \cdot a = (b \cdot a) + (c \cdot a)$ for all $a, b, c \in R$

Rings that also satisfy commutativity for multiplication are called commutative rings.

Definition II.2. A semiring is a set R equipped with two binary operations + and · satisfying the following axioms:

1) $(R, +)$ is a commutative monoid with identity element 0

2) $(R, \cdot)$ is a monoid with identity element 1

3) Multiplication is distributive with respect to addition

4) Multiplication by 0 annihilates R: $0 \cdot a = a \cdot 0 = 0$

The last axiom 4 is omitted from the definition of a ring: it follows from the other ring axioms.

The difference between rings and semirings is that in semirings addition yields only a commutative monoid, not necessary an Abelian group.

An idempotent semiring is one whose addition is idempotent: $a + a = a$.

In the sequel, we assume that all rings are commutative.

Definition II.3. Let R be a commutative ring with operations + and ·
A nonvoid subset I of R is called an ideal of R when

1) a $\in I$ and $b \in I$ imply $a \pm b \in I$

2) For every $r \in R$ and every $a \in I$, $r \cdot a \in I$.

The zero ideal 0 and R itself are also ideals which are trivial ideals.
$(I, +)$ is a subgroup of $(R, +)$.

A principal ideal of a commutative ring R is an ideal generated by one element:

$$(x) = \{xr \mid x \in R \text{ is fixed}, r \in R\}$$

A sum and product of ideals are defined as follows: If I_1 and I_2 are ideals of a ring R then

$$
\begin{aligned}
I_1 + I_2 &= \{a + b \mid a \in I_1 \text{ and } b \in I_2\} \\
I_1 I_2 &= \{a_1 b_1 + \cdots + a_n b_n \mid a_i \in I_1 \text{ and } b_i \in I_2\} \\
i &= 1, \cdots n, \text{ for } n \in \mathbb{N}
\end{aligned}
$$

The sum and product of two ideals are also ideals.

The intersection of two ideals is also an ideal with the operation meet $\wedge$, $I_1 \wedge I_2 = I_1 \cap I_2$.

The join $\vee$ of two ideals is defined as a sum of these ideals. So, $I_1 \vee I_2 = I_1 + I_2$.

The union of two ideals is not necessary an ideal.

The product of $I_1 I_2$ is is contained in the intersection of I_1 and I_2, $I_1 I_2 \subseteq I_1 \cap I_2$.

Let R be a ring and I of ideal of R. We may define an equation relation $\sim$ on R as follows:

$$a \sim b \quad \text{iff} \quad a - b \in I$$

In case $a \sim b$, we say that a and b are modulo I. The equivalence class (the coset) of the element a in R is given by

$$[a] = a + I = \{a + r \mid r \in I\}$$

The set of equivalence classes is denoted by R/I. The operations in R/I are defined as follows:

$$
\begin{aligned}
(a + I) \boxplus (b + I) &= (a + b) + I \\
(a + I) \boxdot (b + I) &= ab + I
\end{aligned}
$$

The zero element of R/I is $0 + I = I$ and the multiplication identity is $1 + I$.

Definition II.4. The set R/I of all equivalence classes (cosets) equipped with operations $\boxplus$, $\boxdot$ form the quotient or factor ring.

We see immediately that if R is a commutative ring, so is R/I. In fact, in this paper we assume that R is commutative. For an Abelian group $(R, +)$ R/I is said to be the quotient group of R.

Let R be a ring with the zero element 0 and the identity 1. If $0 = 1$, then $R = \{0\}$ (with operations) is called the zero ring.

Definition II.5. An integral domain is a commutative ring in which the following cancellation law holds:

If $c \neq 0$ and $ca = cb$, then $a = b$.

Definition II.6. A field is a commutative ring which contains for each nonzero element $a \neq 0$ an inverse element a^{-1} satisfying the equation $a^{-1}a = 1$.

All fields are integral domains. We need later the following result:

Proposition II.7. [2] If R is a commutative ring (with identity) then R is a field iff it has no non-trivial ideals.

This means that the only ideals of R are the zero ideal and R itself.

Proof: Assume that R is a field, and we show that it has no non-trivial ideals. It is sufficient to show that if I is a nonzero ideal in R, then $I = R$. Since I is nonzero, it contains some element $a \neq 0$ R is a field, so it has the inverse a^{-1} and I contains $1 = aa^{-1}$. Since I contains 1, it also contains $x = x1$ for every $x \in R$ and then $I = R$. Conversely, we assume that R has no non-trivial ideals, and we prove that R is a field by showing that if x is a nonzero element in R, then it has the inverse. The set $I = \{yx \mid y \in R\}$ of all multiple of x by elements of R is an ideal. Since I contains $x = 1x$, it has a nonzero ideal, and consequently equals R. We conclude from this that I contains 1, and therefore that there is an element y in R such that $yx = 1$. This shows that x has an inverse, so R is a field. ∎

B. Łukasiewicz rings and ideals

In this subsection we refer to [1]
Let R be a commutative ring which satisfies the condition:

$$\text{for all } x \in R \text{ there is } r \in R \text{ such that } xr = x. \qquad (1)$$

Let $Id(R)$ be the set of all ideals of R. Especially, R itself belongs to $Id(R)$. If operations $+$ and $\cdot$ are the ideal sum and product, respectively, and if 0 is the zero ideal and 1 the ideal R, we set $Sem(R) = (Id(R), +, \cdot, 0, 1)$. Then Sem(R) is a semiring in which R = 1 will act as an identity element. Sem(R) is an idempotent semiring with identity and moreover for ideals I and J we have that $I \subseteq J$ iff $I + J = J$.

Define on Sem(R) the operation $\star : Sem(R) \longrightarrow Sem(R)$ by setting $I^\star = ann(I)$, the annihilator of I, $I^\star = \{x \in R \mid yx = 0 \text{ for all } y \in I\}$.

We are ready to define a Łukasiewicz ring in the following way:

Definition II.8. A commutative ring R is a Łukasiewicz ring provided that

(a) R satisfies condition (1) above.

(b) $I + J = (I^\star \cdot (I^\star \cdot J)^\star)^\star$ for all $I, J \in Id(R)$.

Proposition II.9. For a Łukasiewicz ring we have

(a) In a Łukasiewicz ring we have $I = I^{\star\star}$ for each ideal I.

(b) Let $R = R_1 \times \cdots \times R_n$. Then any finite product of Łukasiewicz rings is a Łukasiewicz ring. Conversely, if a product of rings is a Łukasiewicz ring, so is each factor.

(c) If I is an ideal of R, we have $I = I_1 \times \cdots \times I_n$ and $I^\star = I_1^\star \times \cdots \times I_n^\star$, for ideals $I_j \subseteq R_j$

(d) $I + J = (I_1 + J_1) \times \cdots \times (I_n + J_n)$

III. Fuzzy sets and fuzzy numbers

In this section we refer to [3] and will only survey triangular and trapezoidal fuzzy numbers because it is enough for our purposes.

Definition III.1. Let X be a set. A membership function μ_A of a fuzzy set $A \subseteq X$ is a function

$$\mu_A : X \longrightarrow [0,1].$$

Therefore every element x of X has a membership degree $\mu_A \in [0,1]$.
Consider a set of pairs $\{(x, \mu A(x)) \mid x \in X\} = B$. For every $x \in X$ we have a number $\mu_A(x) \in [0,1]$. Observe that two different x may have the same membership degree. Instead of A we can consider B.
For example, a fuzzy set $A =' $ x is about $6'$ can be represented in different ways but one of them is

$$\tilde{6} = (3;0,1) + (6;1,0) + (9;0,1)$$

if

$$\mu_{\tilde{6}} = \frac{1}{1 + (x-6)^2}$$

where + means the set theoretic union. In general for pairs

$$(x_1; \mu_A(x_1)), (x_2; \mu_A(x_2)), \cdots, (x_n; \mu_A(x_n))$$

we can think that a fuzzy set

$$A = (x_1; \mu_A(x_1)) + (x_2; \mu_A(x_2)) + \cdots + (x_n; \mu_A(x_n))$$

although A is not a subset of X. It is provided that the original A is finite or countable.

Let $A, B \subseteq X$ be fuzzy sets. The membership functions of $A \cup B$ and $A \cap B$ are defined as follows:

$$\mu_{A \cup B}(x) = \max(\mu_A(x), \mu_B(x))$$

and

$$\mu_{A \cap B}(x) = \min(\mu_A(x), \mu_B(x)), \ x \in X \qquad (2)$$

We need these definitions later but not represented by Łukasiewitz t-norm and t-conorm.

Definition III.2. A fuzzy number A is a normal convex fuzzy set of real numbers $\mathbb{R}$ whose membership function μ_A is a piecewise continuous function such that if $a < b < c < d$ are real numbers , the following conditions hold:

(a) $\mu_A(x) = 0$ for every $x \in]-\infty, a[\cup]d, \infty[$.

(b) $\mu_A(x)$ is increasing if $x \in [a, b]$ and decreasing if $x \in [c, d]$.

(c) $\mu_A(a) = 0$, $\mu_A(d) = 0$ and $\mu_A(x) = 1$ for every $x \in [b, c]$.

This definition is sufficient for triangular and trapezoidal fuzzy numbers.

Example III.3.

$$\mathcal{A}(x) = \begin{cases} 0 & x < a_1 \\ \frac{x-a_1}{a_2-a_1} & a_1 \leq x < a_2 \\ 1 & a_2 \leq x < a_3 \\ \frac{a_4-x}{a_4-a_3} & a_3 \leq x \leq a_4 \\ 0 & x > a_4 \end{cases}$$

is a trapezoidal fuzzy number if $a_2 \neq a_3$ but a triangular fuzzy number if $a_2 = a_3$. In the first case we denote $\mathcal{A} = (a_1, a_2, a_3, a_4)$. In the second case

$$\mathcal{A}(x) = \begin{cases} 0 & x < a_1 \\ \frac{x-a_1}{a_2-a_1} & a_1 \leq x < a_2 \\ 1 & x = a_2 = a_3 \\ \frac{a_4-x}{a_4-a_3} & a_3 \leq x \leq a_4 \\ 0 & x > a_4 \end{cases}$$

IV. Connection between fuzzy numbers and ideals of Łukasiewicz rings

Corollary IV.1. Real numbers $\mathbb{R}$ and the quotient ring $\mathbb{R}/I$ are Łukasiewicz rings.

Proof: Because $\mathbb{R}$ and $\mathbb{R}/I$ are fields they have only the trivial ideals (Proposition II.7), the zero ideal and the whole space. Therefore the condition (b) in Definition II.8, $I + J = (I^\star \cdot (I^\star \cdot J)^\star)^\star$ is satisfied and easy to check. Also the condition (a) is satisfied. ■

Figure 1. Lines $y = kx + c$ (corresponding to polynomials) intersect at the same point $x = a$. By Corollary IV.2 below any line is a linear combination of any two other lines. Fixing these lines they form 'a basis'.

Corollary IV.2. In Figure 1, any line $y = kx+c$ (a polynomial $kx+c$) can be represented as a linear combination of two given lines $y = k_1x + c_1$ and $y = k_2x + c_2$ (polynomials $k_1x + c_1$ and $k_2x + c_2$).

In fact, polynomials need not have the same value at the point x, or all the lines need not intersect at x.

Proof:

$$\begin{aligned} a(k_1x + c_1) + b(k_2x + c_2) &= kx + c \\ (ak_1 + bk_2)x + (ac_1 + bc_2) &= kx + c \end{aligned}$$

Then

$$\begin{aligned} ak_1 + bk_2 &= k \\ ac_1 + bc_2 &= c \end{aligned}$$

and

$$a = \frac{kc_2 - k_2c}{k_1c_2 - k_2c_1}$$
$$b = \frac{-kc_1 - k_1c}{k_1c_2 - k_2c_1}$$

∎

Corollary IV.3. Let

$$\mathcal{A} = (a_1, a_2, \cdots a_n) \text{ be a fuzzy number and}$$
$$R_{\mathcal{A}} = \mathbb{R}/(a_1) \times \cdots \times \mathbb{R}/(a_n)$$

There is one - one correspondence between the fuzzy number $\mathcal{A}$ and the ideal $Id(R_{\mathcal{A}})$.

Proof: Corollary IV.1 implies that $\mathbb{R}/(a_i)$ is a Lukasiewicz ring and so the conclusions for it in Proposition II.9 are valid. Let

$$\mathcal{A} = (a_1, a_2, \cdots, a_n) \text{ be a fuzzy number and}$$
$$R_{\mathcal{A}} = \mathbb{R}/(a_1) \times \cdots \times \mathbb{R}/(a_n)$$
$$Id(R_{\mathcal{A}}) = Id(\mathbb{R}/(a_i)) \times \cdots \times Id(\mathbb{R}/(a_n))$$

Since $\mathbb{R}/(a_i)$ is a field, by Proposition II.7, the only ideals of $\mathbb{R}/(a_i)$ are the zero ideal and $\mathbb{R}/(a_i)$ itself. It follows that

$$Id(\mathbb{R}/(a_i)) = \{c/(a_i) \mid c \in \mathbb{R}\} = \{c + (a_i) \mid c \in \mathbb{R}\} =$$
$$= \{ka_i + c \mid \text{ for every } c \in \mathbb{R}, \text{ every } k \in \mathbb{R}$$
$$\text{and a fixed } a_i \in \mathbb{R}\}$$

Every ideal $Id(\mathbb{R}/(a_i))$ can be thought to consists of polynomials $kx + c$ which have some values (not necessary the same) at the point a_i. By Corollary IV.2 such a polynomial $kx + c$ can be represented as a linear combination of other two polynomials $k_1x + c_1$ and $k_2x + c_2$. We choose these polynomials ($k_1x + c_1$ and $k_2x + c_2$) such that they have the same value at the point a_i. All the lines corresponding to all the polynomials which have the same value at the point a_i intersect exactly at this point a_i (see Figure 1). Because (in our case) every none zero ideal of $\mathbb{R}/(a_i)$ is $\mathbb{R}/(a_i)$ itself, we see that geometrically $\mathcal{A} = (a_1, a_2, \cdots, a_n)$ is corresponding to the direct product $R_{\mathcal{A}} = \mathbb{R}/(a_1) \times \cdots \times \mathbb{R}/(a_n) = Id(\mathbb{R}/(a_1)) \times \cdots \times Id(\mathbb{R}/(a_n))$ such that for every ideal $Id(\mathbb{R}/(a_i))$ we can choose any two lines from different sides of the vertical line $x = a_i$ to be 'a basis'. Let $y = k_1x + c_1$ and $y = k_2x + c_2$ these basic elements. Figure 2 describes how some $\mathcal{A}$ corresponds to different parts $\mathbb{R}/(a_i)$. Altogether, we have proved one to one correspondence between the fuzzy number $\mathcal{A}$ and the ideal $Id(R_{\mathcal{A}})$. ∎

Figure 2. A geometrical representation a fuzzy number $\mathcal{A} = (a_1, a_2, \cdots, a_5)$. The four lines $y = k_i + c_i$, $i = 1, 2, 3, 4$

form pairwise (three pairs) 'basic element pairs' for the ideals $Id(\mathbb{R}/(a_i))$, $i = 2, 3, 4$. In every pair the elements intersect at the points a_i, $i = 2, 3, 4$. All the basic elements form a broken line which describes a membership function of $\mathcal{A}$. In this figure we have a purely triangular fuzzy number case.

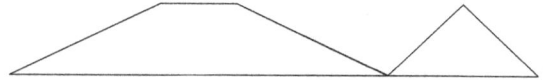

Figure 3. In Figure 3 we have another fuzzy number $\mathcal{B}$.

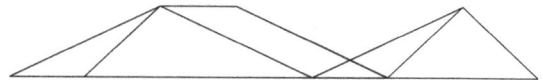

Figure 4. A geometrical representation of a fuzzy number $\mathcal{A} \cup \mathcal{B}$ where $\mathcal{A} = (a_1, a_2, \cdots, a_5)$ and $\mathcal{B} = (b_1, b_2, \cdots, b_6)$ and $a_1 < \cdots < a_5$, $b_1 < \cdots < b_6$. A membership function of $\mathcal{A} \cup \mathcal{B}$ is easy to sketch because $\mu_{\mathcal{A}}(x) \cup \mu_{\mathcal{B}}(x) = \max\{\mathcal{A}(x), \mathcal{B}(x)\}$. The points where the membership functions of $\mathcal{A}$ and $\mathcal{B}$ intersect are $b_1, a_2 = b_2, c, a_4 = b_5, a_5 = b_6$ where $a_3 < c < b_4$. We ignore the intersecting points which are less than b_1 and greater than $a_5 = b_6$.

Next we consider such fuzzy numbers $\mathcal{C}$ whose membership functions intersect at least one point. Denote by $\overline{F}$ the set of such fuzzy numbers $\mathcal{C}$ and let $c_1, \cdots, c_k$ be the common points where the membership functions intersect. We can take only a finite number of common points $c_1 < c_2 < \cdots < c_k$ ignoring all the common points (infinite number) before c_1 and after c_k. In Figure 4 there are the two above fuzzy numbers $\mathcal{A}$ and $\mathcal{B}$ (being in Figures 2 and 3) in the same figure. Their membership functions intersect at five points. $\overline{F}$ consists of fuzzy numbers $\mathcal{C}$ and we set

$$\overline{R}_{\mathcal{C}} = \mathbb{R}/(c_1) \times \cdots \times \mathbb{R}/(c_k)$$
$$Id(\overline{R}_{\mathcal{C}}) = Id(\mathbb{R}/(c_1)) \times \cdots \times Id(\mathbb{R}/(c_k))$$

$$Id(\mathbb{R}/(c_i)) = \{c/(c_i) \mid c \in \mathbb{R}\} = \{c + (c_i) \mid c \in \mathbb{R}\} =$$
$$= \{kc_i + c \mid \text{ for every } c \in \mathbb{R}, \text{ every } k \in \mathbb{R}$$
$$\text{and a fixed } c_i \in \mathbb{R}\}$$

Let $\mathcal{A} = (a_1, a_2, \cdots, a_n)$ and $\mathcal{B} = (b_1, b_2, \cdots, b_m)$. Although they are not necessary the $\mathcal{A}$ and the $\mathcal{B}$ described in Figure 4 we visualize the situation by means of Figure 4.

. Let $c_i \neq a_1, \cdots, c_n, b_1, \cdots, b_m$. We can find the nearest points of c_i: $a_{ni} \in \{a_1 \cdots a_n\}$ and $b_{mi} \in \{b_1 \cdots b_m\}$ such that $a_{ni} < c_i < b_{mi}$ or $b_{mi} < c_i < a_{ni}$.

Especially, by Equation (2) on page 3 the membership function of $\mathcal{A} \cup \mathcal{B}$ is defined by

$$\mu_{A \cup B}(x) = \max\{\mu_A(x), \mu_B(x)\}$$

So, for a membership function of $\mathcal{A} \cup \mathcal{B}$ we can assume that

$$\mu_{\mathcal{A} \cup \mathcal{B}}(x) = \max\{k_{1i}x + d_{ni}, k_{2i}x + d_{mi}\}$$

when $a_{ni} < x < b_{mi}$ or $b_{mi} < x < a_{ni}$. If c_i is some common point of the membership functions $\mathcal{A}$ and $\mathcal{B}$ it leads about the same conclusion.

Altogether, we summarize

$$Id(\mathbb{R}/(c_i)) = Id(\mathbb{R}/(a_{ni})) \vee Id(\mathbb{R}/(b_{mi}))$$

By Corollary IV.3, there is one to one correspondence between the set $\overline{F}$ of fuzzy numbers $\mathcal{C}$ and the set $\overline{G}$ of ideals $Id(\overline{R}_\mathcal{C})$. Therefore,

Proposition IV.4. The mapping

$$h : (\overline{F}, \cup, \cap) \longrightarrow (\overline{G}, \vee, \wedge) \quad h(\mathcal{C}) = Id(\overline{R}_\mathcal{C})$$

is a lattice isomorphism between lattices $(\overline{F}, \cup, \cap)$ and $(\overline{G}, \vee, \wedge)$.

Proof:

From the above discussion and since the sum of ideals is the join of them, and using Proposition II.9 (d) we obtain

$$h(\mathcal{A} \cup \mathcal{B}) = Id(\overline{R}_{\mathcal{A} \cup \mathcal{B}})$$

$$= (Id(\mathbb{R})/(a_{n1})) \vee (Id(\mathbb{R}/(b_{m1})) \times \cdots \times (Id(\mathbb{R}/(a_{nk})) \vee (Id(\mathbb{R}/(b_{mk}))$$

$$= (Id(\mathbb{R}/(a_{n1})) + (Id(\mathbb{R}/(b_{m_1})) \times \cdots \times (Id(\mathbb{R}/(a_{nk})) + (Id(\mathbb{R}/(b_{mk}))$$

$$= (Id(\mathbb{R}/(a_{n1}) \times \cdots \times Id(\mathbb{R}/(a_{nk})) + (Id(\mathbb{R}/(b_{m1}) \times \cdots \times Id(\mathbb{R}/(b_{mk}))$$

$$= Id(\overline{R}_\mathcal{A}) + Id(\overline{R}_\mathcal{B}) = Id(\overline{R}_\mathcal{A}) \vee Id(\overline{R}_\mathcal{B}) = h(\mathcal{A}) \vee h(\mathcal{B})$$

Because

$$\mu_{A \cap B}(x) = \min(\mu_A(x), \mu_B(x))$$

and it is known that the intersection of ideals is the meet of them we have

$$h(\mathcal{A} \cap \mathcal{B}) = Id(\overline{R}_{\mathcal{A} \cap \mathcal{B}})$$

$$= (Id(\mathbb{R})/(a_{n1})) \wedge (Id(\mathbb{R}/(b_{m1})) \times \cdots \times (Id(\mathbb{R}/(a_{nk})) \wedge (Id(\mathbb{R}/(b_{mk}))$$

$$= (Id(\mathbb{R}/(a_{n1})) \cap (Id(\mathbb{R}/(b_{m1}))) \times \cdots \times (Id(\mathbb{R}/(a_{nk})) \cap (Id(\mathbb{R}/(b_{mk}))$$

$$= (Id(\mathbb{R}/(a_{n1}) \times \cdots \times Id(\mathbb{R}/(a_{nk})) \cap (Id(\mathbb{R}/(b_{m1} \times \cdots \times Id(\mathbb{R}/(b_{mk}))$$

$$= Id(\overline{R}_\mathcal{A}) \cap Id(\overline{R}_\mathcal{B}) = Id(\overline{R}_\mathcal{A}) \wedge Id(\overline{R}_\mathcal{B}) = h(\mathcal{A}) \wedge h(\mathcal{B})$$

∎

V. CONCLUSIONS

Geometrically the main result is based on lines and therefore we restricted ourselves to such fuzzy numbers which can be constructed by lines. From this it arises a question that is it possible to create an isomorphism between more general fuzzy numbers and some kind of ideals. Moreover, L.P. Belluce and A. Di Nola formed a connection between Łukasiewicz rings and a certain kind of MV-algebras. But in our case the MV-algebra reduces to a trivial Boolean algebra. So, what does MV-algebras seem if we it is possible to consider more general fuzzy numbers.

REFERENCES

[1] Lawrence P. Belluce and Antonio Di Nola (2009). Commutative rings whose ideals form an MV-algebra. In *Math.Log.Quart. 55, No.5, 468-483 (2009) / DOI 10.1002/malq.200810012*

[2] Garrett Birkhoff and Saunders MacLane (1972). A survey of Modern Algebra. In *Macmillan Publishing Co. (1972). Printed in the United States of America*

[3] Robert Fuller (2000). Introduction to Neuro-Fuzzy Systems. In *Physica-Verlag Heidelberg 1988. Printed in Germany*

Transitions between fuzzified aggregation operators

(Extended Abstract)

Pavel Holeček, Jana Talašová

Dept. of Mathematical Analysis and Applications of Mathematics
Palacký University Olomouc
Olomouc, Czech Republic
pavel.holecek@upol.cz, jana.talasova@upol.cz

Abstract—In the multiple-criteria decision-making, it is often necessary to combine evaluations according to different criteria into a single one. Many aggregation operators can be used for this task (e.g. weighted average, Choquet integral, etc.). When the evaluations are expressed by fuzzy numbers, fuzzified versions of the well-known aggregation operators are used. However, there is usually a trade-off between the flexibility of the aggregation operator (e.g. the type of criteria interactions that it can handle) and the number of parameters that have to be set. This can present a problem when the number of criteria is high. In this paper, two methods that can help overcoming the problem will be presented. They are based on the following general idea – instead of using a complex aggregation method directly, the evaluation model is being built progressively. First, a simple model (using some of the supported aggregation methods) that represents only a rough approximation is designed. This model is then refined and the simple aggregation method is replaced with a more complex one. The parameters of the original method are used for deriving parameters for the more complex one automatically, so that their results would be as close as possible. In the paper, two methods for this task will be discussed. The first one is intended for the fuzzified Choquet integral, while the second one creates fuzzy rule base for a fuzzy expert system.

Keywords—fuzzy; multiple-criteria decision-making; Choquet integral; fuzzy expert system

I. Introduction

In the multiple-criteria decision-making (MCDM), it is necessary to combine evaluations according to the individual criteria into an overall evaluation. This task can be accomplished by some of the aggregation operators.

Probably, the best-known and the most frequently used aggregation method is the weighted average. The expert has to specify a weight for each criterion, which is a relatively simple task. Specifically, for n criteria, $n-1$ parameters have to be provided (the last weight is uniquely determined so that the total sum would be 1).

In 1988, Yager [10] introduced the ordered weighted average (OWA). The expert sets weights, which are not however linked to the individual criteria as in the case of the weighted average, but to the aggregated values according to their ordering. Again, as in case of the weighted average, for n criteria, $n-1$ parameters are required.

A combination of the previous two approaches can be achieved by the weighted OWA (WOWA) introduced by Torra [9]. Two vectors of weights have to be provided – the first one is connected to the criteria as in case of the weighted average, the second one to the evaluations according to their ordering as in OWA. For n criteria, $2(n-1)$ parameters have to be set.

The Choquet integral [2] represents an even more advanced aggregation operator. The Choquet integral requires a fuzzy measure to be defined. For this aggregation operator, it is typical that it can handle also special kinds of criteria interactions (*complementarity*, or *redundancy*). This advantage is, however, counterweighted by the number of parameter that has to be provided by the expert – 2^n-2 values of the fuzzy measure have to be set, which can be quite a difficult task. The problem can be partially reduced by use of k-additive fuzzy measures that require lower amount of parameters while they are still usable for most of the practical problems.

In fuzzy MCDM, the evaluations according to the criteria are represented mostly by fuzzy numbers. To obtain the overall fuzzy evaluation, fuzzified versions of the mentioned aggregation operators can be used. In this paper, we will consider the fuzzy weighted average defined in [7], fuzzy OWA operator proposed in [8], fuzzified WOWA [5], and fuzzy Choquet integral [1], which uses a FNV-fuzzy measure (fuzzy number valued fuzzy measure) [1]. Their area of applications remains the same but the aggregated evaluations and their parameters are represented with fuzzy numbers (with a sole exception of fuzzified WOWA, which requires the weight to be real numbers).

Instead of fuzzifications of classical aggregation operators, a fuzzy expert system can be also used for the aggregation. Its great advantage is that it can be used even when there are complex relationships among the criteria. The expert is required to set a fuzzy rule base. This way, any Borel measurable function can be approximated by a fuzzy expert system and a suitable inference algorithm with an arbitrary precision [6]. The number of rules depends on the required quality of such an approximation. Generally speaking, in fuzzy expert systems, even more parameters are usually used then in case of the previously mentioned fuzzified aggregation operators.

From the practical point of view, it can be very difficult for the expert to set all values of the FNV-fuzzy measure for the

Proceedings of NSAIS16 - 2016 Lappeenranta Finland - ISBN 978-952-265-986-6

fuzzified Choquet integral correctly, especially if the number of criteria is high. Similarly, in case of the fuzzy expert system, it can be a time-consuming task to design the desired fuzzy rule base. This paper focuses on two methods that can simplify these tasks significantly. First ideas have been presented in [4]. This paper extends those ideas significantly and provides corresponding examples of the usage.

II. DESIGNING FUZZY MCDM MODELS USING TRANSITIONS BETWEEN FUZZIFIED AGGREGATION OPERATORS

When a complex fuzzy MCDM model should be created, sometimes it can be advantageous to start with a much simpler model first. The model represents only a very rough approximation of the reality. It can be based, for instance, on the fuzzy weighted average or fuzzy OWA. Next, the simple aggregation method in this initial model is replaced by a more complex one (e.g. the fuzzified Choquet integral). The methods that will be discussed in this paper proposes the parameters (e.g. FNV-fuzzy measure) for the new more complex method automatically, so that the evaluation results would be as close as possible to the results obtained with the original simple model. Finally, the expert modifies only those parameters, which requires an adjustment. This can be much simpler and quicker than the case when all the parameters would be set directly. The advantages of this approach will be demonstrated on examples. In this paper, following two methods will be discussed – the first one uses the fuzzified Choquet integral, the latter one is intended for a fuzzy expert system.

A. Transition to the fuzzified Choquet integral

The first method assumes that the fuzzy weighted average, fuzzy OWA, or fuzzified WOWA is used in the original simple model and that the expert would like to use the fuzzified Choquet integral instead. The method proposes corresponding FNV-fuzzy measure for this purpose.

B. Transition to the fuzzy expert system

In the second method, the fuzzy weighted average, fuzzy OWA, fuzzzified WOWA, or fuzzified Choquet integral can be used in the original simple model. The method proposes corresponding fuzzy rule base for the fuzzy expert system.

III. PRACTICAL USAGE OF THE DESCRIBED THEORY IN THE FUZZME SOFTWARE

The FuzzME [5] is a software tool that makes it possible to design complex fuzzy MCDM models. All aggregation methods mentioned in this paper are supported by the software. The whole theory described in this paper has been implemented in the software, too. It is thus possible to study the behavior of the described methods in practice easily. The demo version of the software is available at http://www.FuzzME.net.

IV. CONCLUSION

Two methods, which can be used when a complex fuzzy MCDM model should be designed, were proposed. In the first one, the fuzzified Choquet integral should be used instead of one of the supported aggregation methods. The corresponding FNV-fuzzy measure is proposed. Similarly, the second method proposes the corresponding fuzzy rule base for a fuzzy expert system. The discussed theory has been implemented into a software tool called FuzzME so that the readers can try it on their own.

ACKNOWLEDGMENT

The research has been supported by the grant GA14-02424S *Methods of operations research for decision support under uncertainty* of the Grant Agency of the Czech Republic.

REFERENCES

[1] I. Bebčáková, J. Talašová, and O. Pavlačka, "Fuzzification of Choquet Integral and its application in multiple criteria decision making," *Neural Network World*, *20*, 2010, pp. 125–137.

[2] G. Choquet, "Theory of capacities," *Annales de L'institut Fourier*, *5*, 1953, pp. 131–295.

[3] M. Grabisch, "k-order additive discrete fuzzy measures and their representation," *Fuzzy Sets and Systems* 92, 1997, pp 167–189.

[4] P. Holeček, and J. Talašová, "Multiple-Criteria Fuzzy Evaluation in FuzzME – Transitions Between Different Aggregation Operators," in J. Talašová, J. Stoklasa, and T. Talášek (Eds.), *Proceedings of the 32nd International Conference on Mathematical Methods in Economics MME 2014*, pp. 305–310, 2014.

[5] P. Holeček, J. Talašová, and I. Müller, "Fuzzy Methods of Multiple-Criteria Evaluation and Their Software Implementation," in V. K. Mago, and N. Bhatia (Eds.), *Cross-Disciplinary Applications of Artificial Intelligence and Pattern Recognition: Advancing Technologies* (pp. 388–411). IGI Global, pp. 388-411, 2012.

[6] B. Kosko, "Fuzzy thinking: The new science of fuzzy logic," New York: Hyperion, 1993.

[7] O. Pavlačka, and J. Talašová, "Application of the Fuzzy Weighted Average of Fuzzy Numbers in Decision-Making Models", New Dimensions in Fuzzy Logic and Related Technologies, Proceedings of the 5th EUSFLAT Conference, Ostrava, Czech Republic, September 11-14 2007, (Eds. M. Štěpnička, V. Novák, U. Bodenhofer), II, 2007, pp. 455–462.

[8] J. Talašová, and I. Bebčáková, "Fuzzification of Aggregation Operators Based on Choquet Integral," *Aplimat - Journal of Applied Mathematics*, *1*(1), 2008, pp. 463–474.

[9] V. Torra, "The weighted OWA operator," *International Journal of Intelligent Systems*, *12*(2), 1997, pp. 153–166.

[10] R.R. Yager, "On ordered weighted averaging aggregation operators in multicriteria decision making," *IEEE Trans.Systems Man Cybernet*, 3 *(1)*, 1988, pp. 183–190.

On Fuzzy-Valued Propositional Logic

Jorma K. Mattila

School of Engineering Science
Lappeenranta University of Technology
P.O.box 20, Fin-53851 Lappeenranta, Finland
Email: jorma.mattila@lut.fi

Abstract—Kleene algebras of fuzzy numbers are considered. Suitable sets of fuzzy numbers are described and examined for presenting the special conditions for those fuzzy numbers which can form a Kleene algebra. A set of triangular fuzzy numbers, called Lambda-standard fuzzy numbers are introduced. Some sets of these fuzzy numbers satisfy the conditions of Kleene algebra. These sets of fuzzy numbers can be used as sets of truth values in Kleene-like fuzzy-valued logics, called *fuzzy-valued many-valued logics*. Some preliminary examples for this kind of logics are sketched. The connection of Łukasiewicz' implication to these logics is shown. This connection gives a possibility to construct some fuzzy-valued Łukasiewicz' logics.

Keywords: ordering of fuzzy numbers, fuzzy-valued Kleene algebras, fuzzy-valued propositional logic

I. BACKGROUND

In this article we consider an algebraic theory for formal propositional n-valued logic ($n \geq 3$) for the case where the truth values are fuzzy numbers or fuzzy intervals. We use the abbreviation FVPL standing for the term *fuzzy-valued propositional logic*.

Fuzzy number-valued Zadeh-algebras (*cf.* Mattila [8]) serve the theoretical basis for the construction of fuzzy-valued propositional logics. The important things are ways of representing, ordering, and forming fuzzy meet and fuzzy join of fuzzy numbers.

A. Kaufmann and M. M. Gupta [4] consider interval arithmetics applied to triangular and trapezoidal fuzzy numbers (or fuzzy intervals) presented by means of α-cuts. They also introduced some criteria for ordering of fuzzy numbers. Besides this, also H. Banderer and S. Gottwald [1] and R. Fullér [3] have considered ordering of fuzzy numbers by defining *fuzzy max* and *fuzzy min* operations by means of Zadeh's extension principle, and V. Novák [9] has defined *join* '$\sqcup$' and *meet* '$\sqcap$' also by means of extension principle, as follows. Let $\mathcal{A}$, $\mathcal{B}$ be fuzzy numbers and $x, y \in \mathbb{R}$. *Join* $\mathcal{A} \sqcup \mathcal{B}$ is the fuzzy number

$$(\mathcal{A} \sqcup \mathcal{B})(z) = \bigvee_{z=x \vee y} (\mathcal{A}(x) \wedge \mathcal{B}(y)). \quad (1)$$

Meet $\mathcal{A} \sqcap \mathcal{B}$ is the fuzzy number

$$(\mathcal{A} \sqcap \mathcal{B})(z) = \bigvee_{z=x \wedge y} (\mathcal{A}(x) \wedge \mathcal{B}(y)). \quad (2)$$

These operations are actually the same as *fuzzy max* and *fuzzy min* in [1] and [3].

Novák [9] also presented the following theorem.

Theorem 1 (Novák). *The fuzzy numbers form a distributive lattice with respect to the operations '$\sqcap$' and '$\sqcup$'. It*

means that

$$\mathcal{A} \sqcup \mathcal{A} = \mathcal{A}$$
$$\mathcal{A} \sqcap \mathcal{A} = \mathcal{A}$$
$$\mathcal{A} \sqcup \mathcal{B} = \mathcal{B} \sqcup \mathcal{A}$$
$$\mathcal{A} \sqcap \mathcal{B} = \mathcal{B} \sqcap \mathcal{A}$$
$$(\mathcal{A} \sqcup \mathcal{B}) \sqcup \mathcal{C} = \mathcal{A} \sqcup (\mathcal{B} \sqcup \mathcal{C})$$
$$(\mathcal{A} \sqcap \mathcal{B}) \sqcap \mathcal{C} = \mathcal{A} \sqcap (\mathcal{B} \sqcap \mathcal{C})$$
$$\mathcal{A} \sqcup (\mathcal{B} \sqcap \mathcal{C}) = (\mathcal{A} \sqcup \mathcal{B}) \sqcap (\mathcal{A} \sqcup \mathcal{C})$$
$$\mathcal{A} \sqcap (\mathcal{B} \sqcup \mathcal{C}) = (\mathcal{A} \sqcap \mathcal{B}) \sqcup (\mathcal{A} \sqcap \mathcal{C})$$

Then he defined the ordering relation $\sqsubseteq$ for fuzzy numbers $\mathcal{A}$, $\mathcal{B}$ in the general way:

$$\mathcal{A} \sqsubseteq \mathcal{B} \quad \text{iff} \quad \mathcal{A} \sqcap \mathcal{B} = \mathcal{A} \quad (\mathcal{A} \sqcup \mathcal{B} = \mathcal{B}). \quad (3)$$

Associative Kleene algebras are involved already in Prof. Zadeh's fuzzy set theory, presented in Zadeh [11], to a great extend. So, these algebras of fuzzy sets are often called *Zadeh algebras*. Fuzzy-valued Zadeh-algebras are associative Kleene algebras of suitable fuzzy quantities, like fuzzy numbers, fuzzy intervals etc. Such a Zadeh-algebra consists of a set of fuzzy quantities satisfying Theorem 1 and some additional properties.

Definition 1. Let T_n be a set $T_n = \{\mathcal{A}_1, \mathcal{A}_2, \ldots, \mathcal{A}_n\}$ of fuzzy quantities, and '$\sqcup$' and '$\sqcap$' are fuzzy join and fuzzy meet operations, respectively. Then

$$\mathcal{T}_n = \langle T_n, \sqcup, \sqcap, \neg, \mathcal{A}_n \rangle$$

is a *fuzzy-valued Zadeh algebra* (*f-v. Z-algebra*, for short) if it satisfies the conditions

(Z1) $\langle T_n, \sqcup, \sqcap \rangle$ is a distributive lattice with respect to '$\sqcap$' and '$\sqcup$';

(Z2)   the neutral element of the operations $\sqcup$ and $\sqcap$ are $\mathcal{A}_1$ and $\mathcal{A}_n$, respectively;

(Z3)   for any $\mathcal{A}_i \in T_n$, there exists $\neg \mathcal{A}_i \in T_n$, such that

$$\neg \mathcal{A}_n \sqsubseteq \neg \mathcal{A}_i \sqsubseteq \neg \mathcal{A}_1;$$

(Z4) $\mathcal{A}_1 \not\models \mathcal{A}_n$,

(Z5)   for any $\mathcal{A}_i, \mathcal{A}_j \in T_n$, $\mathcal{A}_i \sqcap \neg \mathcal{A}_i \sqsubseteq \mathcal{A}_j \sqcup \neg \mathcal{A}_j$.

The condition (Z1) tells that $\langle T_n, \sqcup, \sqcap \rangle$ satisfies the conditions of Theorem 1, especially, the operations $\sqcup$ and $\sqcap$ are commutative, associative, and distributive on T_n. The condition (Z2) means that for all $\mathcal{A}_i \in T_n$, $\mathcal{A}_i \sqcup \mathcal{A}_1 = \mathcal{A}_i$ and $\mathcal{A}_i \sqcap \mathcal{A}_n = \mathcal{A}_i$. The condition (Z3) means that the unary operation '$\neg$' is order reversing on T_n. The condition (Z5) tells that Z-algebra for fuzzy quantities fulfills the so-called *Kleene condition*. Further, if $\langle T_n, \sqcup, \sqcap, \neg, \mathcal{A}_n \rangle$ is a fuzzy-valued Z-algebra then the top element of T_n is $\mathcal{A}_n$. Its negation $\neg \mathcal{A}_n$, i.e., $\mathcal{A}_1$, is the bottom element of T_n.

In general, Kleene algebra is DeMorgan algebra where the Kleene condition holds. So, Zadeh algebra is a special case of Kleene algebra.

The ordering defined in (3) is very important. There exist a lot of cases where using the formulas (1) and (2), i.e., calculating by using extension principle, the results, say, $\mathcal{A} \sqcup \mathcal{B} = \mathcal{B}$ or $\mathcal{A} \sqcap \mathcal{B} = \mathcal{A}$ do not hold. The result may be something like a fuzzy number $\mathcal{C}$ having some parts from both fuzzy numbers $\mathcal{A}$ and $\mathcal{B}$. In the end of Example 2, such a case is considered. Hence, when we are choosing fuzzy quantities to the set T_n we must take care that the ordering condition (3) holds.

II. SPECIAL PROPERTIES FOR APPLICABLE FUZZY QUANTITIES

Examples about easily manipulable fuzzy quantities in applications based on f.-v. Z-algebras of fuzzy sets are triangular fuzzy numbers, Gaussian fuzzy numbers, other bell-shaped fuzzy numbers and fuzzy intervals. In this paper we mainly consider triangular fuzzy numbers. Some considerations are done for trapezoidal fuzzy intervals in Mattila [8] and some other considerations of fuzzy numbers already earlier (*cf.* e.g. in Mattila [5], [6]).

In general, the membership function of a triangular fuzzy number has the form

$$
\mathcal{A}(x) = \Lambda(x; a_1, a_2, a_3) = \begin{cases} 0 & x < a_1 \\ \frac{x - a_1}{a_2 - a_1} & a_1 \leq x < a_2 \\ \frac{a_3 - x}{a_3 - a_2} & a_2 \leq x \leq a_3 \\ 0 & x > a_3 \end{cases} \quad (4)
$$

where $x, a_1, a_2, a_3 \in \mathbb{R}$.

Its shape is a triangle (indicated by the symbol Λ), and the coordinates of the apexes of the triangle are $(a_1, 0)$, $(a_2, 1)$, and $(a_3, 0)$. The support of $\mathcal{A}(x)$ is the interval $[a_1, a_3]$ and the core is the point a_2. This means that $\mathcal{A}(a_1) = 0$, $\mathcal{A}(a_2) = 1$, and $\mathcal{A}(a_3) = 0$.

We adopt the symbol $\tilde{\mathbb{R}}_\Lambda$ standing for the set of all real triangular fuzzy numbers. Hence, the fuzzy number $\Lambda(x; a_1, a_2, a_3)$ described in (4) belongs to the set $\tilde{\mathbb{R}}_\Lambda$.

Usually, considerations are done in a closed real interval, say, $[0, p]$ $(0 < p, p \in \mathbb{R})$, and further, $x, a_1, a_2, a_3 \in [0, p]$ and $0 \leq a_1 < a_2 < a_3$. This interval $[0, p]$ is called *operative interval*.

Further, a *complementarity* for $\mathcal{A}$ on the interval $[0, p]$ can be given in the similar form

$$
\neg \mathcal{A}(x) = \Lambda(x; p - a_3, p - a_2, p - a_1)
$$

$$
= \begin{cases} 0 & x < p - a_3 \\ \frac{x - p + a_3}{a_3 - a_2} & p - a_3 \leq x < p - a_2 \\ \frac{p - a_1 - x}{a_2 - a_1} & p - a_2 \leq x \leq p - a_1 \\ 0 & x > p - a_1 \end{cases} \quad (5)
$$

which is a triangular fuzzy number on $[0, p]$. For this complementarity we use the name *negation of $\mathcal{A}(x)$ on the interval $[0, p]$*.

By means of the construction of the fuzzy number $\mathcal{A}(x) = \Lambda(x; a_1, a_2, a_3)$ in (4) and its negation $\neg \mathcal{A}(x) = \Lambda(x; p - a_3, p - a_2, p - a_1)$ in (5), the both triangles, with the apexes $(a_1, 0)$, $(a_2, 1)$, $(a_3, 0)$ and $(p - a_3, 0)$, $(p - a_2, 1)$, $(p - a_1, 0)$, have the same area, because the lengths of their supports equal to $a_3 - a_1$ and they have the same height $\mathcal{A}(a_2) = \mathcal{A}(p - a_2) = 1$. So, the both areas are $A = \frac{1}{2}(a_3 - a_1)$.

In addition to this, if the arithmetic mean of a_1 and a_3 happens to be a_2 the both triangles have exactly the same shape, and they are isosceles triangles, i.e., the line segments between the points $(a_1, 0)$, $(a_2, 1)$ and $(a_2, 1)$, $(a_3, 0)$ have the equal length. Fuzzy numbers having the same shape are called *similar*.

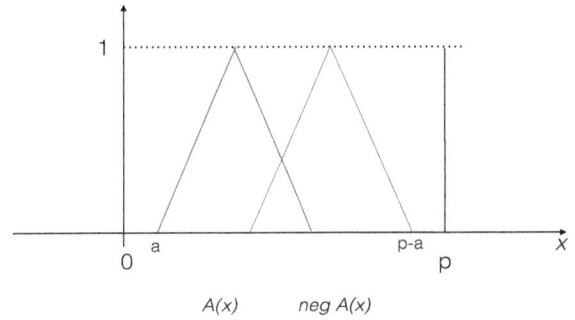

Fig. 1. Λ-shaped fuzzy number and its negation

Kaufmann and Gupta [4] have considered *linear ordering* of fuzzy numbers. They give three criteria for determining the order of given fuzzy numbers to be applied in the following order. Let $\mathcal{A}_i(x) = \Lambda(x; a_i, b_i, c_i)$ $(i = 1, \ldots, k)$ be Λ-shaped fuzzy numbers. The criteria are as follows:

(i) *Removal criterion*: For each $\mathcal{A}_i$, calculate the removal value

$$
\widehat{\mathcal{A}}_i = \frac{a_i + 2b_i + c_i}{4} \quad (6)
$$

The fuzzy number having the biggest removal value is the biggest one. If the result of this step is not unique then continue.

(ii) *Mode criterion*: Compare the cores $\text{core}\mathcal{A}_j = b_j$ of those fuzzy numbers having the biggest removal value. The fuzzy number having the biggest core is the biggest fuzzy number. If the result of this step is not unique then continue.

(iii) *Divergence criterion*: Determine the length of the each $\text{supp}\mathcal{A}_l$ that are still left. The fuzzy number having the biggest support is the biggest fuzzy number.

Example 1. Let $\mathcal{B}(x) = \Lambda(x; b_1, b_2, b_3) \in \tilde{\mathbb{R}}_\Lambda$ be another real triangular fuzzy number. Compare the fuzzy numbers $\mathcal{A}(x) = \Lambda(x; a_1, a_2, a_3)$ (see the formula (4)) and $\mathcal{B}(x)$. Applying the formula (6) we have the removal values

$$\widehat{\mathcal{A}} = \frac{a_1 + 2a_2 + a_3}{4} \quad \text{and} \quad \widehat{\mathcal{B}} = \frac{b_1 + 2b_2 + b_3}{4} \quad (7)$$

Consider the situations where $\mathcal{A}(x)$ is bigger than $\mathcal{B}(x)$, i.e., $\mathcal{A}(x) \sqsupset \mathcal{B}(x)$. According to the removal criterion, if $\widehat{\mathcal{A}} < \widehat{\mathcal{B}}$ then $\mathcal{A}(x) \sqsubset \mathcal{B}(x)$. If this is not the case, we have to apply the mode criterion. If the cores satisfy the condition $a_2 < b_2$ then $\mathcal{A}(x) \sqsubset \mathcal{B}(x)$. If this is not the case, we have to apply the divergence criterion. The lengths of the supports of $\mathcal{A}(x)$ and $\mathcal{B}(x)$ are $a_3 - a_1$ and $b_3 - b_1$, respectively. If $a_3 - a_1 < b_3 - b_1$ then $\mathcal{A}(x) \sqsubset \mathcal{B}(x)$. According to the formula (3), we have the case $\mathcal{A}(x) \sqcup \mathcal{B}(x) = \mathcal{B}(x)$. Hence, the case $\mathcal{A}(x) \sqcap \mathcal{B}(x) = \mathcal{A}(x)$ holds, too.

Example 2. Calculate fuzzy min and fuzzy max of the fuzzy numbers $\mathcal{A}(x) = \Lambda(x; 1, 5, 6)$ and $\mathcal{B}(x) = \Lambda(x; 2, 4, 7)$.

Removal criterion:

$$\widehat{\mathcal{A}} = \frac{1 + 2 \times 5 + 6}{4} = \frac{17}{4} \text{ and } \widehat{\mathcal{B}} = \frac{2 + 2 \times 4 + 7}{4} = \frac{17}{4} \quad (8)$$

Because the removal values are equal, the removal criterion does not give any difference. Hence we need to apply the next criterion.

Mode criterion:

The cores are $\operatorname{core}\mathcal{A} = 5$ and $\operatorname{core}\mathcal{B} = 4$. So, $\operatorname{core}\mathcal{B} < \operatorname{core}\mathcal{A}$, and hence $\mathcal{B}(x) \sqsubset \mathcal{A}(x)$.

Because the mode criterion gives the difference between $\mathcal{A}$ and $\mathcal{B}$, we need not to use the divergence criterion.

The extension principle does not give this result. When applying the formulas (1) and (2) to these fuzzy numbers we get

$$(\mathcal{A} \sqcup \mathcal{B})(x) = \begin{cases} \mathcal{A}(x) & \text{if } 3 < x \leq 5.5 \\ \mathcal{B}(x) & \text{if } 2 \leq x \leq 3 \text{ or } 5.5 \leq x \leq 7 \end{cases} \quad (9)$$

and

$$(\mathcal{A} \sqcap \mathcal{B})(x) = \begin{cases} \mathcal{A}(x) & \text{if } 1 \leq x \leq 3 \text{ or } 5.5 \leq x \leq 6 \\ \mathcal{B}(x) & \text{if } 3 < x \leq 5.5 \end{cases} \quad (10)$$

Let us have a short consideration about trapezoidal fuzzy interval. The general form of the membership function of a trapezoidal fuzzy interval is

$$\mathcal{A}(x) = \Gamma(x; a_1, a_2, a_3, a_4) = \begin{cases} 0 & \text{if } x < a_1 \\ \frac{x - a_1}{a_2 - a_1} & \text{if } a_1 \leq x < a_2 \\ 1 & \text{if } a_2 \leq x < a_3 \\ \frac{a_4 - x}{a_4 - a_3} & \text{if } a_3 \leq x \leq a_4 \\ 0 & \text{if } x > a_4 \end{cases} \quad (11)$$

on a closed real interval $[0, p]$ $(0 < p, \ p \in \mathbb{R})$ and $x, a_1, a_2, a_3, a_4 \in [0, p]$, and $0 \leq a_1 < a_2 < a_3 < a_4$. If $a_2 = a_3$ then $\mathcal{A}$ is a trapezoidal fuzzy interval. The apexes of the triangular fuzzy interval (11) are $(a_1, 0)$, $(a_2, 1)$, $(a_3, 1)$, and $(a_4, 0)$. The support and the core of $\mathcal{A}$ are the intervals $\operatorname{supp}\mathcal{A} = [a_1, a_4]$ and $\operatorname{core}\mathcal{A} = [a_2, a_3]$.

The negation of a trapezoidal fuzzy interval $\mathcal{A}(x) = \Gamma(x; a_1, a_2, a_3, a_4)$ in a given interval $[0, p]$ can be calculated in the similar way as that of a triangular fuzzy number. Hence, The negation of $\mathcal{A}(x)$ on the interval $[0, p]$ is $\neg\mathcal{A}(x) = \Gamma(x; p - a_4, p - a_3, p - a_2, p - a_1)$.

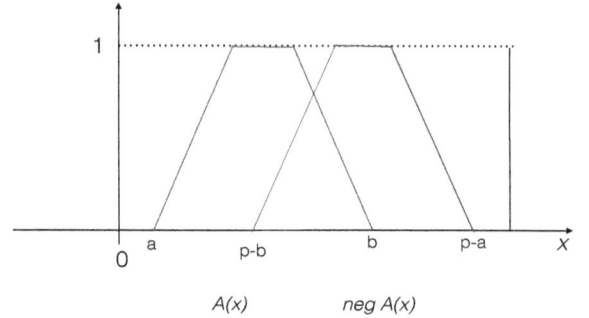

Fig. 2. Trapezoidal fuzzy numbers

For ranking trapezoidal fuzzy intervals, the above given three ordering criteria can be applied. The only separation is in the removal criterion. For a trapezoidal interval $\mathcal{A}(x) = \Gamma(x; a_1, a_2, a_3, a_4)$ the removal criterion is

$$\widehat{\mathcal{A}} = \frac{a_1 + a_2 + a_3 + a_4}{4} \quad (12)$$

Hence, triangular fuzzy numbers and trapezoidal fuzzy intervals are comparable, and they can be ordered into linear order.

III. FUZZY NUMBERS FOR TRUTH VALUE SYSTEMS

Consider a set of n similar fuzzy triangular numbers (or alternatively, similar trapezoidal intervals) of equal size $T_n = \{\mathcal{A}_0, \mathcal{A}_1, \ldots, \mathcal{A}_{n-1}\}$ on the interval $[0, p]$, such that

$$[0, p] \subset \bigcup_{i=0}^{n-1} \operatorname{supp}\mathcal{A}_i \quad (13)$$

holds. This is done such that the first non-decreasing half of $\operatorname{supp}\mathcal{A}_0$ and the second non-increasing half of $\operatorname{supp}\mathcal{A}_{n-1}$ leave out of the interval $[0, p]$. So, the cores of $\mathcal{A}_0$ and $\mathcal{A}_{n-1}$ are $\operatorname{core}\mathcal{A}_0 = 0$ and $\operatorname{core}\mathcal{A}_{n-1} = p$, i.e., the end points of the interval $[0, p]$. The reason for this construction is that the $\mathcal{A}_0$ represent the truth value *false* and $\mathcal{A}_{n-1}$ the truth value *true*, and the full falsehood is situated at the point $x = 0$ and it degrases when x increases, and the full truth is situated at the point $x = p$ and it increases from 0 to 1 when x approaches p from the left in the first half of $\operatorname{supp}\mathcal{A}_{n-1}$. See the figure of truth values in Sec. 5.

We want to construct the set T_n such that it is a finite universe of a fuzzy-valued Zadeh algebra. The following consideration holds in the both cases, i.e., for the interval $[0, p]$ (where $\mathcal{A}_0$ and $\mathcal{A}_{n-1}$ are truncated) and for the extended interval where the whole supports of fuzzy numbers $\mathcal{A}_0$ and $\mathcal{A}_{n-1}$ are included in the interval.

Proceedings of NSAIS16 - 2016 Lappeenranta Finland - ISBN 978-952-265-986-6

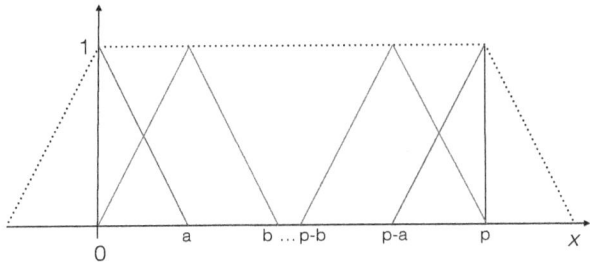

Fig. 3. Fuzzy truth values

It can be easily seen that the conditions of Theorem 1 hold. So,

$$\mathcal{T}_n = \langle T_n, \sqcup, \sqcap \rangle \tag{14}$$

is a distributive lattice, by Theorem 1. Hence, the condition (Z1) of Definition 1 holds.

The condition (Z2) holds by construction.

We construct the set of n fuzzy numbers T_n in such a way that if a fuzzy number $\mathcal{A}_i$ belongs to T_n then $\neg\mathcal{A}_i$ belongs to T_n, by the definition of negation which is given by the formula (5). This means that for any $x \in [0,p]$, the curve of the union of the fuzzy set $\mathcal{A}(x) \vee \neg\mathcal{A}(x)$ is symmetric with respect to the vertical line $x = \frac{p}{2}$ at the mid point $\frac{p}{2}$ of the interval $[0,p]$. For example, check the fuzzy numbers $\mathcal{A}$ and $\neg\mathcal{A}$ in the formulas (4) and (5). Hence we see that $\neg\mathcal{A}_0 = \mathcal{A}_{n-1}$, $\neg\mathcal{A}_1 = \mathcal{A}_{n-2}$, ... , $\neg\mathcal{A}_{n-2} = \mathcal{A}_1$, $\neg\mathcal{A}_{n-1} = \mathcal{A}_0$ which can be given as a general formula

$$\neg\mathcal{A}_k = \mathcal{A}_{n-k-1} \tag{15}$$

Hence, the set T_n is closed with respect to negation.

The condition (Z4) holds by construction.

At last, we show that the system $\mathcal{T}_n = \langle T_n, \sqcup, \sqcap, \neg, \mathcal{A}_n \rangle$ satisfies the Kleene condition (Z5).

Let $\mathcal{A}_i, \mathcal{A}_j \in T_n$ be arbitrarily chosen, hence $\neg\mathcal{A}_i, \neg\mathcal{A}_j \in T_n$, too, because T_n is closed under negation. Suppose $\mathcal{A}_i \sqsubseteq \mathcal{A}_j$ whenever $i \leq j$, for all $\mathcal{A}_i, \mathcal{A}_j \in T_n$.

If the number of fuzzy sets in T_n is $n = 2k+1$ (i.e., n is odd) then the middle element of T_n is $\mathcal{A}_{k+1}$, and hence $\neg\mathcal{A}_{k+1} = \mathcal{A}_{k+1}$, by the definition of negation.

If $n = 2k$ (i.e., n is even) then $\neg\mathcal{A}_k = \mathcal{A}_{k+1}$ and $\neg\mathcal{A}_{k+1} = \mathcal{A}_k$, by the definition of negation.

We denote

$$\mathcal{A}_{[\frac{n}{2}]} = \begin{cases} \mathcal{A}_{k+1} & \text{if } n = 2k+1 \\ \mathcal{A}_k & \text{if } n = 2k \end{cases}$$

Hence, for any $\mathcal{A}_i$, if $\mathcal{A}_i \sqsubseteq \mathcal{A}_{[\frac{n}{2}]}$ then $\mathcal{A}_{[\frac{n}{2}]} \sqsubseteq \neg\mathcal{A}_i$, and vice versa. Hence, for any i, $\mathcal{A}_i \sqcap \neg\mathcal{A}_i \sqsubseteq \mathcal{A}_{[\frac{n}{2}]}$ and for any j, $\mathcal{A}_{[\frac{n}{2}]} \sqsubseteq \mathcal{A}_j \sqcup \neg\mathcal{A}_j$.

So, the system $\mathcal{T}_n = \langle T_n, \sqcup, \sqcap, \neg, \mathcal{A}_n \rangle$ is a fuzzy-valued Z-algebra.

The lattice $\langle T_n, \sqcap, \sqcup \rangle$ is also complete because for any two elements $\mathcal{A}_j, \mathcal{A}_k \in T_n$ ($j,k \leq n$), the expressions $\mathcal{A}_j \sqcup \mathcal{A}_k$ and $\mathcal{A}_j \sqcap \mathcal{A}_k$ are defined and $\mathcal{A}_j \sqcup \mathcal{A}_k, \mathcal{A}_j \sqcap \mathcal{A}_k \in T_n$.

As an additional case, we consider *De Morgan Laws* on the set T_n. The order of the elements of T_n is now linear, and we know which one of these is the negation of any given element of T_n.

Double negation law

$$\neg\neg\mathcal{A}_j = \mathcal{A}_j \tag{16}$$

holds on T_n. The result follows by (15).

Let $\mathcal{A}_j, \mathcal{A}_k \in T_n$ be any two elements and $\mathcal{A}_j \sqsubseteq \mathcal{A}_k$. Hence, it is easy to see that $\neg\mathcal{A}_k \sqsubseteq \neg\mathcal{A}_j$. From this it follows that $\neg\mathcal{A}_k \sqcup \neg\mathcal{A}_j = \neg\mathcal{A}_j$ and hence $\neg(\neg\mathcal{A}_k \sqcup \neg\mathcal{A}_j) = \neg\neg\mathcal{A}_j = \mathcal{A}_j = \mathcal{A}_j \sqcap \mathcal{A}_k$, because $\mathcal{A}_j \sqcap \mathcal{A}_k = \mathcal{A}_j$, by the assumption, i.e., $\mathcal{A}_j \sqcap \mathcal{A}_k = \neg(\neg\mathcal{A}_k \sqcup \neg\mathcal{A}_j)$. The case $\neg(\neg\mathcal{A}_k \sqcap \neg\mathcal{A}_j) = \mathcal{A}_k \sqcup \mathcal{A}_j$ follows from the first case by replacing $\mathcal{A}_j$ and $\mathcal{A}_k$ with $\neg\mathcal{A}_j$ and $\neg\mathcal{A}_k$, respectively. Hence, De Morgan Laws

$$\mathcal{A}_j \sqcap \mathcal{A}_k = \neg(\neg\mathcal{A}_k \sqcup \neg\mathcal{A}_j), \tag{17}$$
$$\mathcal{A}_j \sqcup \mathcal{A}_k = \neg(\neg\mathcal{A}_k \sqcap \neg\mathcal{A}_j)$$

hold on the set T_n.

For fuzzy truth values, we have a lot of possibilities to choose the style of fuzzy numbers to represent truth values. Here we choose a set of some special subset of Λ-numbers of the set $\tilde{\mathbb{R}}_\Lambda$, called Λ-*standard fuzzy numbers*. Before giving the definition for these numbers, we define another concept, called *speciality index* given by R. Yager.[1]

Definition 2. Let $\mathcal{A}$ be a fuzzy subset of $\mathbb{R}$, then the kth *speciality index* of $\mathcal{A}$ is a real number

$$i_k(\mathcal{A}) = \frac{1}{\int_a^b \mathcal{A}^k(x)dx}, \quad k = 0,1,2,\dots \tag{18}$$

where the interval $[a,b]$ is the support of $\mathcal{A}$.

Note that if $k = 1$ then the integral $\int_a^b \mathcal{A}^k(x)dx$ gives the area bordered by the x-axis and the curve of $\mathcal{A}(x)$. If $k = 0$ then the integral gives the length of $\text{supp}\mathcal{A}(x)$.

Definition 3. A normal fuzzy subset $\mathcal{A}$ of $\mathbb{R}$ is a Λ-*standard fuzzy number* if and only if

(i) The membership function of $\mathcal{A}$ is of the form (4), i.e.,

$$\mathcal{A}(x) = \Lambda(x; a, n, b), \quad a,n,b \in \mathbb{R}.$$

(ii) The curve of $\Lambda(x; a, n, b)$ is symmetric with respect to the line $x = n$.

(iii) The first speciality index of $\mathcal{A}$ is $i_1(\mathcal{A}) = 1$.

The following theorem is a straight consequence of Definition 3.

Theorem 2. *$\mathcal{A}$ is a Λ-standard real fuzzy number if and only if it is of the form*

$$\mathcal{A}(x) = \Lambda(x; n-1, n, n+1)$$
$$= \begin{cases} x - (n-1) & \text{if } n-1 < x \leq n \\ n+1-x & \text{if } n < x < n+1 \\ 0 & \text{if } \text{either } x \leq n-1 \text{ or } x \geq n+1 \end{cases}$$

The proof is given in Mattila [6].

Proceedings of NSAIS16 - 2016 Lappeenranta Finland - ISBN 978-952-265-986-6

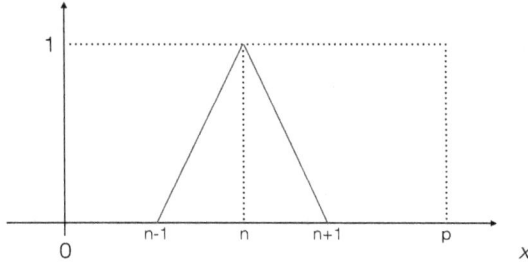

Fig. 4. Λ-standard fuzzy number

If $n \in \mathbb{N}$, we have a Λ-*standard fuzzy integer* $\tilde{n} = \Lambda(x; n-1, n, n+1) \in \tilde{\mathbb{N}}$ where $\tilde{\mathbb{N}}$ is the set of Λ-shaped fuzzy integers.

In the sequel we need the operations *sum* and *difference* of Λ-standard fuzzy numbers. These operations are defined as follows.

Definition 4. Let $\tilde{a}$ and $\tilde{b}$ be any Λ-standard fuzzy numbers. Their *sum* and *difference* are defined as

$$(\tilde{a} \oplus \tilde{b})(x) = \Lambda(x; a+b-1, a+b, a+b+1)$$
$$(\tilde{a} \ominus \tilde{b})(x) = \Lambda(x; a-b-1, a-b, a-b+1)$$

According to Def. 4, we have the result

Theorem 3. *For any Λ-standard fuzzy numbers $\tilde{a}$, $\tilde{b} \in \tilde{\mathbb{R}}_\Lambda$ there is a Λ-standard fuzzy number $\tilde{c} \in \tilde{\mathbb{R}}_\Lambda$, such that*

$$\tilde{c} = \tilde{a} \oplus \tilde{b} \quad iff \quad c = a+b \qquad (addition) \qquad (19)$$
$$\tilde{c} = \tilde{a} \ominus \tilde{b} \quad iff \quad c = a-b \qquad (subtraction) \qquad (20)$$

where $c = \text{core}(\tilde{c})$, $a = \text{core}(\tilde{a})$, *and* $b = \text{core}(\tilde{b})$.

The result follows from the properties of Λ-standard fuzzy numbers.

This kind of fuzzy numbers are very simple to use. Mathematics using Λ-standard fuzzy integers is easy, and most operations are concentrated to manipulations with cores, and hence returned to operations with usual natural numbers. The ordering relation given by the condition (3) works between these numbers using extension principle as well as Kaufmann's and Gupta's method described above.

IV. ESSENTIAL PRELIMINARY THINGS FOR FUZZY-VALUED PROPOSITIONAL LOGIC

We use the abbreviation "FVPL" for fuzzy-valued propositional logic being a general expression for this kind of logical systems.

In general, the set of truth values T_n consists of n Λ-standard integers as follows:

$$T_n = \{\tilde{0}, \tilde{1}, \ldots, \widetilde{n-1}\} \qquad (21)$$

[1]Yager, "Decision with Usual Values", in: Jones et al (eds.), *Fuzzy Sets Theory and Applications*, NATO ASI Series, Publisher Reidal, 1986

where $\tilde{0} = \Lambda(x; -1, 0, 1)$, $\tilde{1} = \Lambda(x; 0, 1, 2)$, $\ldots$, $\widetilde{n-1} = \Lambda(x; n-2, n-1, n)$. So these fuzzy numbers are labelled according to their cores. In this case, the interval $[0, p]$ is $[0, n-1]$ which causes that the first half from $\tilde{0}$ (i.e., the interval $[-1, 0[$) and the second half from $\widetilde{n-1}$ (i.e., the interval $]n-1, n]$ are left out from the operative interval $[0, n-1]$. The reason for this is that the first half of $\tilde{0}$ on the interval $[-1, 0[$ does not actually present the fuzzy truth value *false*, and the second half of $\widetilde{n-1}$ on the interval $]n-1, n]$ does not actually present the fuzzy truth value *true* in our system, so, the *actual truth values false*, i.e., $\tilde{0}$, and *true*, i.e., $\widetilde{n-1}$, have the membership functions

$$\tilde{0}(x) = \begin{cases} 1-x & \text{if} \quad 0 \le x \le 1 \\ 0 & \text{if} \quad 1 < x \le n-1 \end{cases} \qquad (22)$$

and

$$\widetilde{(n-1)}(x) = \begin{cases} 0 & \text{if} \quad 0 \le x < n-2 \\ x-n+2 & \text{if} \quad n-2 \le x \le n-1 \end{cases} \qquad (23)$$

We could have chosen these kind of fuzzy numbers already in the in the benining, because it would not have caused any difference to the theory, especially because the complementarity works, i.e., $\neg\tilde{0} = \widetilde{n-1}$ and $\neg\widetilde{(n-1)} = \tilde{0}$, and the linear order holds.

These Λ-standard fuzzy numbers represent truth values as follows: $\tilde{0} := false$ and $\widetilde{n-1} := true$, the others represent different degrees of truth.

The language of FVPL is a standard propositional language, i.e., it is, in practice, the same as classical propositional language. However, FVPL is nonclassical because of the formal interpretation of logical connectives and the number of truth values. In addition to this, we can assign fuzzy truth values to statements which are really fuzzy.

We define the alphabet of FVPL as follows.

Definition 5. The *alphabet* of FVPL consists of the following symbols:
1) *propositional variables*: $p, p_0, p_1, \ldots, p_k, \ldots$ (either with or without subindex);
2) *primitive connectives*:
   - negation symbol $\neg$,
   - disjunction symbol $\sqcup$,
3) *parentheses* (and).

Propositional variables are usually called *atoms*.

The negation symbol operates in the set of truth values T_n as we saw above. Because T_n forms a f-v. Z-algebra, all the properties of negation of f-v. Z-algebra hold in the set of truth values of these fuzzy-valued logics.

Using the primitive connectives we can create some derived connectives, like *conjunction symbol* $\sqcap$, *implication symbol* $\rightarrow$, and *equivalence symbol* $\leftrightarrow$. We emphasize that they are metasymbols. Another set of metasymbols is any set of capital letters A, B, C, etc. (either with or without subindices) to refer to any formulas of FVPL. We return to these derived connectives after the next definition.

We give the recursive definition of *well formed formulas* of FVPL as follows.

Definition 6. The *well formed formulas* (wff's, for short) of FVPL are

1) propositional variables $p, p_0, p_1, \ldots, p_k, \ldots$ are wff's;
2) if A and B are wff's then
  - $\neg A$ is a wff,
  - $A \sqcup B$ is a wff.

All the wff's of FVPL can be formed using this procedure.

Now, we adopt the *conjunction* $\sqcap$, which actually is an abbreviation of the righthand side of the following expression, as follows.

$$A \sqcap B \overset{\text{def}}{=} \neg(\neg A \sqcup \neg B) \tag{24}$$

This definition is based on DeMorgan's laws (17).

There are many ways for introducing an implication operation for FVPL by means of the primitive connectives. We choose an implication which is an extension of classical material implication. This is a basic case of *S-implications*. We define an S-implication in FVPL by means of negation and disjunction of FVPL as follows:

$$A \rightarrow B \overset{\text{def}}{=} \neg A \sqcup B \tag{25}$$

This implication is a type of *Kleene-Dienes implication* applied to fuzzy-valued truth values. It can be obtained some applications based on FVPL with this implication.

Now, the equivalence symbol can be derived by means of implication in the natural way:

$$A \leftrightarrow B \overset{\text{def}}{=} (A \rightarrow B) \sqcap (B \rightarrow A) \tag{26}$$

A truth value of conjunction $A \sqcap B$ is the truth value of the fuzzy minimum of the components A and B.

A truth value of disjunction $A \sqcup B$ is the truth value of the fuzzy maximum of the components A and B.

A truth value of Kleene-Dienes type implication $A \rightarrow B$ is the fuzzy maximum of of the truth values of $\neg A$ and B.

A truth value of equivalence $A \leftrightarrow B$ is the truth value of those of fuzzy minimum of $A \rightarrow B$ and $B \rightarrow A$.

In the language of FVPL the alphabet (see Definition 5), metavariables, and primitive and derived connectives are often called symbols. When connectives are connected to formulas, they are called also operations.

There is also a connection between FVPL disjunction and Łukasiewicz implication. In original Łukasiewicz' logics where the set of truth values is $[0,1]$ (or its suitable subset including 0 and 1), the calculation formula for this Ł-implication is

$$a \rightarrow b = \min\{1, 1 - a + b\}, \tag{27}$$

where a and b are numerical truth values of a Łukasiewicz' logic. The corresponding calculation formula in FVPL using the truth value set (21), where $\tilde{a}, \tilde{b} \in T_n$, is

$$\tilde{a} \rightarrow \tilde{b} = \tilde{k} \sqcap (\tilde{k} \ominus \tilde{a} \oplus \tilde{b}) \tag{28}$$

where we write $\tilde{k} = \widetilde{n-1}$, for convenience. The symbols '$\oplus$' and '$\ominus$' are arithmetical operations on $\tilde{\mathbb{R}}_\Lambda$, according to Definition 4. It is also easy to see that the following theorem is true.

Theorem 4. *Let* $\tilde{a}, \tilde{b}, \tilde{c}, \tilde{d} \in \tilde{\mathbb{R}}_\Lambda$ *with the corresponding cores* a, b, c, d, *then*

$$\tilde{a} \sqcap \tilde{b} = \tilde{c} \iff \min\{a, b\} = c$$
$$\tilde{a} \sqcup \tilde{b} = \tilde{d} \iff \max\{a, b\} = d$$

Because all the calculations can be done using the cores, by Theorems 3 and 4, we practically use the corresponding formula for the cores $\text{core}(\tilde{a}) = a$, $\text{core}(\tilde{b}) = b$, and $\text{core}(\tilde{k}) = k$ instead of the formula (28) as follows:

$$a \rightarrow b = \min\{k, k - a + b\} \tag{29}$$

The connection between Z-algebra and Łukasiewicz' logic is closely considered in Mattila [7], where numerical many-valued truth values are used.

In the proof of the following theorem we need the useful formulas for the operations max and min. Let a, b be arbitrary real numbers. Then

$$\max\{a, b\} = \frac{a + b + |a - b|}{2} \tag{30}$$
$$\min\{a, b\} = \frac{a + b - |a - b|}{2}$$

We now give the following

Theorem 5. *Let* $\tilde{p}, \tilde{q} \in T_n$ *be any* Λ*-standard fuzzy integers, then*

$$\tilde{p} \sqcup \tilde{q} = (\tilde{p} \rightarrow \tilde{q}) \rightarrow \tilde{q} \tag{31}$$

where the implication is Łukasiewicz implication (28).

Proof: We can use the corresponding cores p, q and k (for $\tilde{k} = \widetilde{n-1}$) of the fuzzy numbers. So, we have to show that $\max\{p, q\} = (p \rightarrow q) \rightarrow q$. We have

$$\max\{p, q\} = \min\{k, \max\{p, q\}\} \qquad p, q \leq k$$
$$= \min\{k, \frac{p + q + |p - q|}{2}\} \qquad \text{by (30)}$$
$$= \min\{k, \frac{2k - 2k + p + q + |k - k + p - q|}{2}\}$$
$$= \min\{k, k - \frac{2k - p - q - |k - (k - p + q)|}{2}\}$$
$$= \min\{k, k - \frac{k + (k - p - q) + 2q - 2q - |k - (k - p + q)|}{2}\}$$
$$= \min\{k, k - \frac{k + (k - p + q) - |k - (k - p + q)|}{2} + q\}$$
$$= \min\{k, k - \min\{k, k - p + q\} + q\} \qquad \text{by(30)}$$
$$= \min\{k, k - (p \rightarrow q) + q\} \qquad \text{by (29)}$$
$$= (p \rightarrow q) \rightarrow q \qquad \text{by (29)}$$

Because the core determines the corresponding Λ-standard fuzzy number, the formula (31) holds, because the corresponding formula for the cores holds. ■

Theorem 5 gives the connection of Łukasiewicz' implication to the other connectives of FVPL. In the formula

(28) arithmetical operations '$\oplus$' and '$\ominus$' for the fuzzy truth values are carried out by using the usual operations '+' and '-' for usual numbers to be applied on the set of the cores of fuzzy truth values, as we did in the formula (29). The core of a fuzzy number $\tilde{a} \to \tilde{b}$, i.e., the fuzzy number $\widetilde{(n-1)} \sqcap \widetilde{(n-1-\tilde{a}+\tilde{b})}$ determines the core of the corresponding fuzzy truth value, which is the value of corresponding Łukasiewicz' fuzzy implication.

Theorem 5 connects the concept of Łukasiewicz' implication to f-v. Z-algebras. However, we use in FVPL the structure of Kleene many-valued logic as a base of our approach to fuzzy-valued logic using Kleene-Dienes type implication.

Kleene-Dienes type implication is very interesting because it is much critical than, for example, Łukasiewicz' implication. One interesting case is that the formula $A \to A$ is not a tautology, i.e., it is not true in all truth value distributions. For example, if $A := \tilde{3}$ when the biggest truth value is, say, 5 representing *true*, we have

$$A \to A := (\neg\tilde{3} \to \tilde{3}) = (\tilde{5} \ominus \tilde{3}) \sqcup \tilde{3} = \tilde{2} \sqcup \tilde{3} = \tilde{3} \neq \tilde{5} \quad (32)$$

by (25). About Kleene many-valued logics, see Bergmann [2] or Rescher [10].

V. EXAMPLE: FIVE-VALUED FVPL

As an example, we consider a five-valued system of fuzzy-valued propositional logic. The truth values are the five Λ-standard fuzzy integers $\tilde{0} = \Lambda(x; -1, 0, 1)$, $\tilde{1} = \Lambda(x; 0, 1, 2)$, $\tilde{2} = \Lambda(x; 1, 2, 3)$, $\tilde{3} = \Lambda(x; 2, 3, 4)$, and $\tilde{4} = \Lambda(x; 3, 4, 5)$, and the operative interval is $[0, 4]$. So, we do not use the left half of $\tilde{0}$ (i.e., the interval $[-1, 0[$) and the right half of $\tilde{4}$ (i.e., the interval $]4, 5]$). The negations of each truth values are as follows:

$$\neg\tilde{0} = \tilde{4}, \ \neg\tilde{1} = \tilde{3}, \ \neg\tilde{2} = \tilde{2}, \ \neg\tilde{3} = \tilde{1}, \ \neg\tilde{4} = \tilde{0}.$$

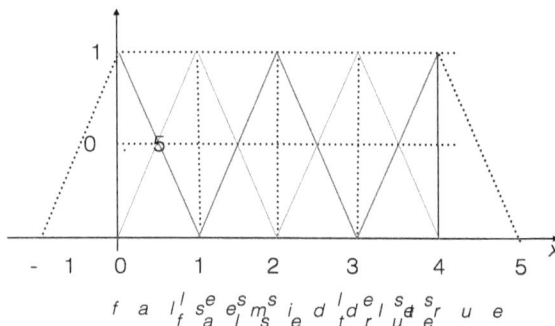

Fig. 5. Five-valued system of truth values

It is easy to see, that the set of truth values

$$T_5 = \{\tilde{0}, \tilde{1}, \tilde{2}, \tilde{3}, \tilde{4}\}$$

forms a complete distributive lattice with respect to the operations $\sqcup$ and $\sqcap$, and it forms a f-v. Z-algebra $\mathcal{T}_5 = \langle T_5, \sqcup, \sqcap, \neg, \tilde{4} \rangle$.

The linguistic labels connected to the truth values can be given for example as follows:

$$\tilde{0} := false, \tilde{1} := less\,false, \tilde{2} := middle,$$

$$\tilde{3} := less\,true, \text{ and } \tilde{4} := true.$$

Example 3. Consider the following statements:
1) If salaries or prices rise, inflation comes.
2) If inflation comes, then the government must control it, or people will suffer.
3) If people will suffer, then the cabinet ministers fall into disgrace.
4) The government does not control inflation, and cabinet ministers do not fall into disgrace.

These statements are *premises* in our logical inference. Does the *conclusion*

- Salaries do not rise.

follow logically from the premises? To solve this problem, we formalize the sentences and consider the truth values for the atoms $S :=$ 'salaries rise', $P :=$ 'prices rise', $I :=$ 'inflation comes', $C :=$ 'the government must control inflation', $F :=$ 'people will suffer', and $D :=$ 'cabin ministers fall into disgrace' in the inference.

The formalized inference is as follows.

$$
\begin{array}{ll}
1. & (S \sqcup P) \to I \\
2. & I \to (C \sqcup F) \\
3. & F \to D \\
4. & \neg C \sqcap \neg D \\
\hline
5. & \neg S
\end{array}
$$

We consider three cases:

(a) What happens when the all premises are true?
(b) What is the truth status if all the premises are not true, using Kleene-Dienes type implication?
(c) What is the truth status if all the premises are not true, using Lukasiewicz' implication?

Case (a): In logical inference the main principle is that *a true conclusion follows from the true premises*.

We examine, wether the conclusion is true, i.e., have the truth value $\neg S := \tilde{4}$, according to this principle. First, we use Kleene-Dienes type implication.

(i) In Step 4., if $\neg C \sqcap \neg D := \tilde{4}$, both $\neg C := \tilde{4}$ and $\neg D := \tilde{4}$, by the property of conjunction. Then $C := \tilde{0}$ and $D := \tilde{0}$, by the property of negation.
(ii) Then in Step 3, $F \to D := \tilde{4}$ because then $F := \tilde{0}$, and $\neg F := \tilde{4}$ and hence $\neg F \sqcup D := \tilde{4}$ by (25). (According to (i), $D := \tilde{0}$.)
(iii) Now in Step 2, $I := \tilde{0}$, because $C := \tilde{0}$ and $F := \tilde{0}$ and hence $C \sqcup F := \tilde{0}$. So, the formula in Step 4 is true.
(iv) Now, in Step 1, $(S \sqcup P) \to I := \tilde{4}$ iff $(S \sqcup P) := \tilde{0}$, because $I := \tilde{0}$. If $S := \tilde{0}$ and $P := \tilde{0}$ then really $(S \sqcup P) := \tilde{0}$ and $(S \sqcup P) \to I := \tilde{4}$.

Proceedings of NSAIS16 - 2016 Lappeenranta Finland - ISBN 978-952-265-986-6

The result of this analysis gives the truth value distribution

$$C := \tilde{0},\ D := \tilde{0},\ F := \tilde{0},\ I := \tilde{0},\ P := \tilde{0}\ S := \tilde{0}, \quad (33)$$

making all the premises true.

The truth value we got for the atomic formula S in the consideration above is $\tilde{0}$, i.e., false. So, its negation $\neg S$ has the truth value $\tilde{4}$, i.e., the conclusion of the inference $\neg S$ is true in that truth value distribution where all the premises are true.

Now, we do the same task using Łukasiewicz, implication. First, we calculate the truth values of premises using the truth value distribution (33). Using the formula (28) we have

1. $(S \sqcup P) \to I := \tilde{0} \sqcup \tilde{0} \to \tilde{0} = \tilde{0} \to \tilde{0} = \tilde{4}$ (34)

2. $I \to (C \sqcup F) := \tilde{0} \sqcup \tilde{0} \to \tilde{0} = \tilde{0} \to \tilde{0} = \tilde{4}$

3. $F \to D := \tilde{0} \to \tilde{0} = \tilde{4}$

4. $\neg C \sqcap \neg D := \tilde{4} \to \tilde{4} = \tilde{4}$

So, the premises are true, and because the conclusion $\neg S$ is true, too, the inference have the truth value *true*.

Case (b): Consider the case where we have the following truth value distribution:

$$S := \tilde{1} \quad \Longleftrightarrow \quad \neg S := \tilde{3} \quad (35)$$
$$P := \tilde{2} \quad \Longleftrightarrow \quad \neg P := \tilde{2}$$
$$I := \tilde{3} \quad \Longleftrightarrow \quad \neg I := \tilde{1}$$
$$F := \tilde{2} \quad \Longleftrightarrow \quad \neg F := \tilde{2}$$
$$C := \tilde{2} \quad \Longleftrightarrow \quad \neg C := \tilde{2}$$
$$D := \tilde{1} \quad \Longleftrightarrow \quad \neg D := \tilde{3}$$

We calculate the truth value of the formula where the conjunction of the premises implies the conclusion, i.e., that of the formula

$$[((S \sqcup P) \to I) \sqcap (I \to (C \sqcup F)) \sqcap (F \to D) \sqcap (\neg C \sqcap \neg D)] \to \neg S \quad (36)$$

Using the formula (25) for Kleene-Dienes type implication we get the formula (36) into the form

$$\neg[(\neg(S \sqcup P) \sqcup I) \sqcap (\neg I \sqcup (C \sqcup F)) \quad (37)$$
$$\sqcap (\neg F \sqcup D) \sqcap (\neg C \sqcap \neg D)] \sqcup \neg S$$

Substituting the given truth values to the atoms and their negations, we have

$$\neg[(\neg(\tilde{1} \sqcup \tilde{2}) \sqcup \tilde{3}) \sqcap (\tilde{1} \sqcup (\tilde{2} \sqcup \tilde{2})) \quad (38)$$
$$\sqcap (\tilde{2} \sqcup \tilde{1}) \sqcap (\tilde{2} \sqcap \tilde{3})] \sqcup \tilde{3}$$
$$= \neg[(\neg\tilde{2} \sqcup \tilde{3}) \sqcap (\tilde{1} \sqcup \tilde{2}) \sqcap (\tilde{2} \sqcup \tilde{2})] \sqcup \tilde{3}$$
$$= \neg[(\tilde{2} \sqcup \tilde{3}) \sqcap \tilde{2} \sqcap \tilde{2}] \sqcup \tilde{3} = \neg(\tilde{3} \sqcup \tilde{2} \sqcup \tilde{2}) \sqcup \tilde{3}$$
$$= \neg\tilde{3} \sqcup \tilde{3} = \tilde{1} \sqcup \tilde{3} = \tilde{3}$$

So, the inference is *less true* if we use the above given truth value distribution. This means that the inference does not always give a totally true conclusion if all the premises are not true.

Case (c): We apply the above given truth value distribution in the case where Łukasiewicz' implication (28)

is used in our inference. In this case the truth value calculation of the formula (36) is as follows.

$$[((\tilde{1} \sqcup \tilde{2}) \to \tilde{3}) \sqcap (\tilde{3} \to (\tilde{2} \sqcup \tilde{2})) \quad (39)$$
$$\sqcap (\tilde{2} \to \tilde{1}) \sqcap (\tilde{2} \sqcap \tilde{3})] \to \tilde{3}$$
$$= [(\tilde{2} \to \tilde{3}) \sqcap ((\tilde{3} \to \tilde{2}) \sqcap (\tilde{2} \to \tilde{1}) \sqcap \tilde{2}] \to \tilde{3}$$
$$= [(\tilde{4} \sqcap (\tilde{4} \ominus \tilde{2} \oplus \tilde{3})) \sqcap (\tilde{4} \sqcap (\tilde{4} \ominus \tilde{3} \oplus \tilde{2}))$$
$$\sqcap (\tilde{4} \sqcap (\tilde{4} \ominus \tilde{2} \oplus \tilde{1})) \sqcap \tilde{2}] \to \tilde{3}$$
$$= [\tilde{4} \sqcap \tilde{3} \sqcap \tilde{3} \sqcap \tilde{2}] \to \tilde{3} = \tilde{2} \to \tilde{3} = \tilde{4} \sqcap (\tilde{4} \ominus \tilde{2} \oplus \tilde{3}) = \tilde{4}$$

In this case the inference is totally *true*.

The Case (a) shows that when the premises are true, the conclusion is true in our example inference in both systems, i.e., using Kleene-Dienes type implication and Łukasiewicz' implication. This case also shows that the inference is correct in "classical" situation when we have only two truth values, *true* and *false*.

In the cases (b) and (c) we used the ruth value distribution (35) consisting of the truth values being not totally true. Kleene-Dienes type implication appeared to more strict than Łukasiewicz' implication. Using Łukasiewicz' implication the inference appeared to be *true* with the truth value distribution (35) but only *less true* using Kleene-Dienes type implication.

VI. Some concluding remarks

The role of associative Kleene algebras, i.e., Z-algebras in fuzzy set theory is quite remarkable. The extension of these algebras from values of fuzzy sets (i.e., from point wise values of membership functions) to some fuzzy quantities, like fuzzy numbers, has appeared to be useful. For example, these fuzzy-valued Z-algebras give a mathematical base for fuzzy screening systems (*cf.* Mattila [8]).

The arithmetical operations on any set of Λ-standard fuzzy numbers we need, are not the same as is usually used in fuzzy numbers. Because every truth value set T_n must be closed with respect to the logical connectives and possibly some other needed operations in manipulations of truth values, these operations must be chosen in a way, that the closeness presupposition is satisfied. In this way, resulting truth values are Λ-standard fuzzy numbers belonging to T_n. We may imagine these special fuzzy numbers in the way that the core of a given triangular fuzzy number is covered by a "constant umbrella", which actually is the curve of the membership function of the fuzzy number. And the core of every Λ-standard fuzzy number in the given truth value set T_n has the "constant umbrella" of the same kind and size. About arithmetic operations for Λ-standard fuzzy numbers, see Mattila [6].

The logical system FVPL we sketched above seems to need some continuation where the fuzzy numbers as truth values would have a deeper role than having linear order. Now we just assign linguistic truth labels to them completely intuitively and subjectively. This assignment is not mathematics any more. The same result can be reached by using Kleene's or Łukasiewicz' "classic" many-valued logic with their usual numeric truth values. In fact, we may need to find a new dimension for the fuzzy truth values. If we observe the definition 3 of Λ-standard numbers closely, we see that any Λ-standard number represents a probability density function, because the area restricted by the curve of Λ-standard number and x-axis is one. But then we need random variables which obey this probability density function. Evidently, this kind of random variables exist. Anyway, this may give a new dimension, but then we must use probability density functions as fuzzy numbers, like suitable Gaussian fuzzy numbers, to be fuzzy truth values. Some questions will rise. What are such situations where FVPL and probabilities can be combined, and how we should do it? All these things are still totally open.

The description of FVPL is here just in the starting stage and not completely defined. The propositional language, connectives with their calculation formulas and connections between them, and truth values are given. About FVPL itself, only some sketches are considered, especially Example 3 gives some hints about the cases which in practice may be included in this kind of logic. So, the work will continue.

REFERENCES

[1] H. Bandemer, S. Gottwald, *Fuzzy Sets, Fuzzy Logic, Fuzzy Methods*, Johm Wiley & Sons, 1995, ISBN 0-471-95636-8.

[2] M. Bergmann, *An Introduction to Many-Valued and Fuzzy Logic*, Cambridge University Press, 2008, ISBN 978-0-521-70757-2.

[3] , R. Fullér, *Introduction to Neuro-Fuzzy Systems*, Advances in Soft Computing, Physica-Verlag, Heidelberg, 2000.

[4] A. Kaufmann, M. M. Gupta, *Fuzzy Mathematical Models in Engineering and Management Science*, North-Holland, Amsterdam, New York, Oxford, Tokyo, 1988.

[5] J. K. Mattila, *A Study on Λ shaped Fuzzy Integers*, Research Report no. 76, Lappeenranta University of Technology, Dept. Information Technology. Lappeenranta, 2000.

[6] J. K. Mattila, On Field Theory of Λ-standard Fuzzy Numbers, in: N. Baba, L.C. Jain, B.J. Howlett (eds.), *Knowledge-Based Intelligent Information Engineering Systems and Allied Technologies*, KES'2001, Frontiers in Artificial Intelligence and Applications, Vol. 69, IOS Press, 2001. p. 695-699.

[7] J. K. Mattila, "Zadeh Algebra as the Basis of Łukasiewicz Logics", in Mark J. Wierman (ed.): *2012 Annual Meeting of the North American Fuzzy Information Processing Society*, August 6-8, Berkeley, CA, IEEE Catalog Number: CFP12750-USB, ISBN: 978-1-4673-2337-6.

[8] J. K. Mattila, "A Note on Fuzzy-Valued Inference", in: M. Collan, M. Fedrizzi, J. Kacprzyk (eds): *Fuzzy Technology*, Studies in Fuzziness and Soft Computing, Vol. 335, Springer International Publishing Switzerland 2016, ISSN 1434-9922, ISBN 078-3-319-26984-9.

[9] V. Novák, *Fuzzy Sets and their Applications*, Adam Hilger, Bristol and Philadelphia, 1989.

[10] N. Rescher, *Many-valued Logic*, McGraw-Hill, 1969.

[11] L. A. Zadeh, Fuzzy Sets, *Information and Control*, 8, 1965.

Linguistic approximation under different distances/similarity measures for fuzzy numbers

Tomáš Talášek
Lappeenranta University of Technology
School of Business and Management
Skinnarilankatu 32, 53851
Lappeenranta, Finland
and
Palacký University
Olomouc, Czech Republic
Email: tomas.talasek@lut.fi

Jan Stoklasa
Lappeenranta University of Technology
School of Business and Management
Skinnarilankatu 32, 53851
Lappeenranta, Finland
Email: jan.stoklasa@lut.fi

Abstract—**The paper explores the performance of two selected distance measures of fuzzy numbers (the Bhattacharyya distance and the dissemblance index) and two different fuzzy similarity measures in the context of linguistic approximation. Symmetrical triangular fuzzy numbers are considered as the approximated entities. A numerical experiment is employed to map the behavior of the four distance/similarity measures under uniform linguistic scales used for the linguistic approximation. We identify a clear difference of the performance of the selected methods when the approximation of low-uncertain fuzzy numbers are approximated. One of the selected fuzzy similarity measures tends to emphasize extreme evaluations and disregard intermediate linguistic values. For highly uncertain fuzzy numbers the methods seem to be close to equivalent in terms of the linguistic approximation.**

Keywords—*Linguistic approximation, fuzzy number, distance, similarity, Bhattacharyya distance.*

I. INTRODUCTION

In real-life applications it is often reasonable to present an output of a mathematical model (e.g. real number, fuzzy set,...) together with a linguistic term, that would describe the output of the model to the decision maker. The users of the outputs of mathematical models are often managers (or even complete laymen in mathematics) who are not sufficiently acquainted with the mathematical model. Description of the results in terms of natural language is a more suitable way to present information to the users of mathematical models [6]. Providing linguistic description of the outputs of mathematical models can increase their credibility, help avoid misinterpretations and can create a creative environment for the user-model interaction. The process of assigning linguistic terms to an output of a mathematical model is called *linguistic approximation*. As the name itself suggests, the result are approximated, which means that a suitable (semantically close) linguistic label is assigned to them. The possible loss or slight distortion of information in the process has to be compensated by a significant increase in understandability of the results, a reduction of time needed to process and understand the results or some other positive effect. Recent research also suggests that the ideas of linguistic approximation can be used e.g. for ordering purposes - see [7]. An appropriate technique of linguistic approximation has to be chosen to achieve these effects. One of the most common approaches to linguistic approximation is based on the distance/similarity measure of fuzzy numbers.

Fuzzy numbers are a useful mathematical representation of the meanings of linguistic labels (elements of natural language). One of the frequently studied classes of fuzzy numbers are symmetrical triangular fuzzy numbers. Their popularity lies in the simplicity of their definition – the decision maker must choose the kernel of fuzzy number (basic value, representative value or typical value) and then the length of the support (uncertainty). Symmetric triangular fuzzy numbers are a frequent building block of linguistic scales - they are used to represent the meanings of its linguistic values. They are frequently used in the evaluation context. On scales with natural minima and maxima (e.g. the interval [0,1] where 0 is the worst possible evaluation and 1 the best possible evaluation), triangular fuzzy numbers can be used to represent the meanings of a predefined set of ordered linguistic evaluation terms, they can even reflect that extreme evaluations tend to be less uncertain (on both poles of the scale) and neutral evaluations tend to be more vague. If the decision maker wants to evaluate alternative as poor (values close to 0) or excellent (values close to 1) he or she must be confident with his opinion and therefore the support of fuzzy number will be small (more certain). However if the evaluation is in the middle, the decision maker may by undecided and therefore the support can by much wider (more uncertain evaluation).

In this paper we investigate role of different distances/similarity measures on the performance of linguistic approximation of symmetric triangular fuzzy numbers. By a numerical study we compare two distances of fuzzy numbers – modified Bhattacharyya distance and dissemblance index together with two fuzzy similarity measures regarding their performance in linguistic approximation using uniform linguistic scales.

II. PRELIMINARIES

Let U be a nonempty set (the universe of discourse). A *fuzzy set* A on U is defined by the mapping $A : U \to [0, 1]$. For each $x \in U$ the value $A(x)$ is called a *membership degree* of the element x in the fuzzy set A and $A(.)$ is called a *membership function* of the fuzzy set A. $\text{Ker}(A) = \{x \in U | A(x) = 1\}$ denotes a *kernel* of A, $A_\alpha = \{x \in U | A(x) \geq \alpha\}$ denotes an α-*cut* of A for any $\alpha \in [0,1]$, $\text{Supp}(A) = \{x \in U | A(x) > 0\}$ denotes a *support* of A. Let A and B be fuzzy sets on the same universe U. We say that A is a *fuzzy subset* of B ($A \subseteq B$), if $A(x) \leq B(x)$ for all $x \in U$.

A fuzzy number is a fuzzy set A on the set of real numbers which satisfies the following conditions: (1) $\text{Ker}(A) \neq \emptyset$ (A is *normal*); (2) A_α are closed intervals for all $\alpha \in (0,1]$ (this implies A is *unimodal*); (3) $\text{Supp}(A)$ is bounded. A family of all fuzzy numbers on U is denoted by $\mathcal{F}_N(U)$. A fuzzy number A is said to be defined on [a,b], if $\text{Supp}(A)$ is a subset of an interval $[a, b]$. Real numbers $a_1 \leq a_2 \leq a_3 \leq a_4$ are called *significant values* of the fuzzy number A if $[a_1, a_4] = \text{Cl}(\text{Supp}(A))$ and $[a_2, a_3] = \text{Ker}(A)$, where $\text{Cl}(\text{Supp}(A))$ denotes a closure of $\text{Supp}(A)$. Each fuzzy number A is determined by $A = \{[\underline{a}(\alpha), \overline{a}(\alpha)]\}_{\alpha \in [0,1]}$, where $\underline{a}(\alpha)$ and $\overline{a}(\alpha)$ is the lower and upper bound of the α-cut of fuzzy number A respectively, $\forall \alpha \in (0,1]$, and the closure of the support of A $\text{Cl}(\text{Supp}(A)) = [\underline{a}(0), \overline{a}(0)]$. A *cardinality* of fuzzy number A on $[a, b]$ is a real number $\text{Card}(A)$ defined as follows: $\text{Card}(A) = \int_a^b A(x)dx$.

The fuzzy number A is called *linear* if its membership function is linear on $[a_1, a_2]$ and $[a_3, a_4]$; for such fuzzy numbers we will use a simplified notation $A = (a_1, a_2, a_3, a_4)$. A linear fuzzy number A is said to be *trapezoidal* if $a_2 \neq a_3$ and *triangular* if $a_2 = a_3$. We will denote triangular fuzzy numbers by ordered triplet $A = (a_1, a_2, a_4)$. Triangular fuzzy number $A = (a_1, a_2, a_4)$ is called *symmetric triangular fuzzy number* if $a_2 - a_1 = a_4 - a_2$. More details on fuzzy numbers and computations with them can be found for example in [2].

In real-life applications we often need to represent fuzzy numbers by real numbers. This proces is called *defuzzification*. The most common method is to substitute fuzzy number by its center of gravity (COG). Let A be a fuzzy number on $[a, b]$ for which $a_1 \neq a_4$. The *center of gravity* of A is defined by the formula $\text{COG}(A) = \int_a^b xA(x)dx / \text{Card}(A)$. If $A = (a_1, a_2, a_4)$ is symmetric triangular fuzzy number on $[a, b]$, then $\text{COG}(A) = \text{Ker}(A) = a_2$.

A *fuzzy scale* on $[a, b]$ is defined as a set of fuzzy numbers $T_1, T_2, \ldots, T_s$ on [a,b], that form a Ruspini fuzzy partition (see [5]) of the interval $[a, b]$, i.e. for all $x \in [a, b]$ it holds that $\sum_{i=1}^s T_i(x) = 1$, and the T's are indexed according to their ordering. A *linguistic variable* ([9]) is defined as a quintuple $(\mathcal{V}, \mathcal{T}(\mathcal{V}), X, G, M)$, where $\mathcal{V}$ is a name of the variable, $\mathcal{T}(\mathcal{V})$ is a set of its linguistic values (terms), X is an universe on which the meanings of the linguistic values are defined, G is an syntactic rule for generating the values of $\mathcal{V}$ and M is a semantic rule which to every linguistic value $\mathcal{A} \in \mathcal{T}(\mathcal{V})$ assigns its meaning $A = M(\mathcal{A})$ which is usually a fuzzy number on X.

III. LINGUISTIC APPROXIMATION OF FUZZY NUMBERS

Let the fuzzy set O on $[a, b]$ be an output of a mathematical model that needs to be linguistically approximated by one linguistic term from the set $\mathcal{T}(\mathcal{V}) = \{\mathcal{T}_1, \ldots, \mathcal{T}_s\}$. $\mathcal{T}(\mathcal{V})$ is a linguistic term set of a linguistic variable $(\mathcal{V}, \mathcal{T}(\mathcal{V}), [a, b], G, M)$,

such that $T_i = M(\mathcal{T}_i)$, $i = 1, \ldots, s$ are fuzzy numbers on $[a, b]$.

Linguistic approximation $\mathcal{T}_O \in \mathcal{T}(\mathcal{V})$ of the fuzzy set O is computed by

$$T_O = \arg \min_{i \in \{1, \ldots, s\}} d(T_i, O) \qquad (1)$$

where $d(A, B)$ is a distance or similarity measure[1] of two fuzzy numbers. During the past forty years a large number of approaches were proposed for computation of distances and similarity of fuzzy numbers (see e.g. [10]). It is necessary to keep in mind that the choice of distance/similarity measure will modify the behavior of the linguistic approximation method. In the next chapter the following distances and similarity measures of fuzzy numbers A and B will be considered:

- *modified Bhattacharyya distance* [1]:

$$d_1(A, B) = \left[1 - \int_U (A^*(x) \cdot B^*(x))^{1/2} dx\right]^{1/2}, \quad (2)$$

where $A^*(x) = A(x)/\text{Card}(A(x))$ and $B^*(x) = B(x)/\text{Card}(B(x))$,

- *dissemblance index* [4]

$$d_2(A, B) = \int_0^1 |\underline{a}(\alpha) - \underline{b}(\alpha)| + |\overline{a}(\alpha) - \overline{b}(\alpha)| \, d\alpha. \quad (3)$$

- *similarity measure* (introduced by Wei and Chen [8])

$$s_1(A, B) = \left(1 - \frac{\sum_{i=1}^4 |a_i - b_i|}{4}\right) \cdot \frac{\min\{Pe(A), Pe(B)\} + 1}{\max\{Pe(A), Pe(B)\} + 1}, \quad (4)$$

where $Pe(A) = \sqrt{(a_1 - a_2)^2 + 1} + \sqrt{(a_3 - a_4)^2 + 1} + (a_3 - a_2) + (a_4 - a_1)$, $Pe(B)$ is defined analogically

- *similarity measure* (introduced by Hejazi and Doostparast [3])

$$s_2(A, B) = \left(1 - \frac{\sum_{i=1}^4 |a_i - b_i|}{4}\right) \cdot \frac{\min\{Pe(A), Pe(B)\}}{\max\{Pe(A), Pe(B)\}} \cdot \frac{\min\{Ar(A), Ar(B)\} + 1}{\max\{Ar(A), Ar(B)\} + 1} \quad (5)$$

where $Ar(A) = \frac{1}{2}(a_3 - a_2 + a_4 - a_1)$, $Ar(B)$ is defined analogically and $Pe(A)$ and $Pe(B)$ are computed identically as in the previous method.

IV. NUMERICAL EXPERIMENT

For the purpose of this paper we restrict ourselves to the linguistic approximation of symmetric triangular fuzzy numbers. We assess the behavior of the distance measures d_1 and d_2 and of the two similarity measures s_1 and s_2 in the process of linguistic approximation method by the following numerical experiment. A five element linguistic scale was defined on the interval $[0, 1]$. The meanings of its linguistic terms $\{\mathcal{T}_1, \ldots, \mathcal{T}_5\}$ were chosen to be triangular fuzzy numbers

---

[1]In the case of similarity measure the *arg min* function in formula (1) must be replaced by *arg max* function.

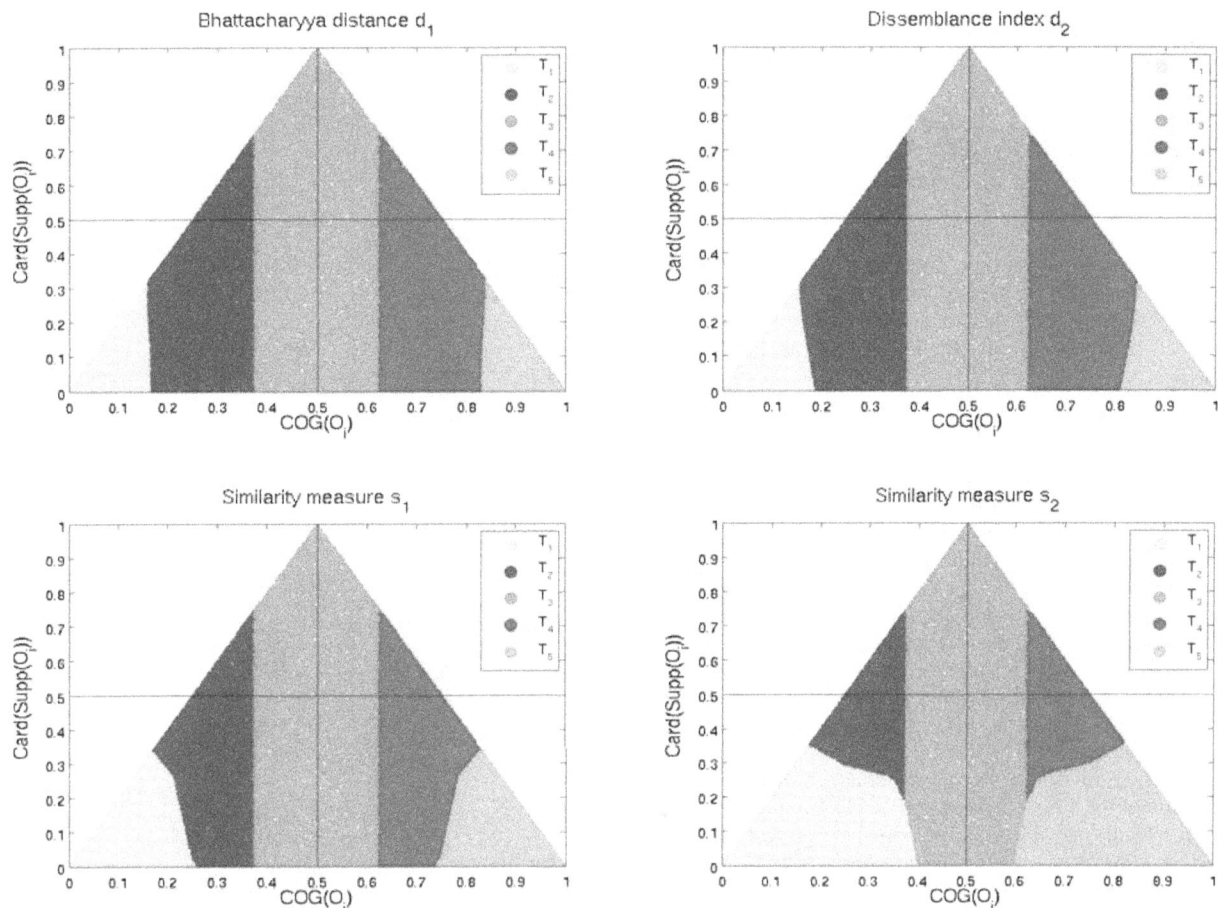

Fig. 1. Results of the numerical experiment. Each randomly generated symmetrical triangular fuzzy number $\{O_1, \ldots, O_{200000}\}$ is represented by a point in two-dimensional space, its coordinates are given by its COG and the cardinality of its Support. The color of the point reflects the resulting linguistic approximation term as suggested by d_1 (top left plot), d_2 (top right plot), s_1 (bottom left plot) and s_2 (bottom right plot). Each plot consists of 200 000 points.

$\{T_1, \ldots, T_5\} = \{(0, 0, 0.25), (0, 0.25, 0.5), (0.25, 0.5, 0.75), (0.5, 0.75, 1), (0.75, 1, 1)\}$ forming a uniform Ruspini fuzzy partition of $[0, 1]$. 200000 symmetric triangular fuzzy numbers $\{O_1, \ldots, O_{200000}\}$ on $[0, 1]$ were randomly generated to be linguistically approximated using this linguistic scale. Each of the four distance and similarity measures (2)–(5) was applied to find the appropriate linguistic approximation for each generated fuzzy number. The sets $\{\mathcal{T}_{O_1}^{d_1}, \ldots, \mathcal{T}_{O_{200000}}^{d_1}\}$, $\{\mathcal{T}_{O_1}^{d_2}, \ldots, \mathcal{T}_{O_{200000}}^{d_2}\}$, $\{\mathcal{T}_{O_1}^{s_1}, \ldots, \mathcal{T}_{O_{200000}}^{s_1}\}$ and $\{\mathcal{T}_{O_1}^{s_2}, \ldots, \mathcal{T}_{O_{200000}}^{s_2}\}$ were thus obtained.

Figure 1 summarizes the results of the numerical experiment graphically. Each plot represents one of the chosen distance or similarity measures and the colours identify the linguistic approximation term suggested by the respective distance/similarity. The generated symmetric triangular fuzzy numbers $\{O_1, \ldots, O_{200000}\}$ are represented by their centers of gravity (horizontal axis) and by the cardinality of their supports (vertical axis).

V. DISCUSSION

It is clear from the plots, that the performance of d_1, d_2, s_1 and s_2 is not identical with respect to the linguistic approx-

imation output (note that a 5-term uniform linguistic scale was applied). It is true to say that for the generated fuzzy numbers with higher uncertainty (cardinality of the support around 0.4 or higher), the performance of the distances and similarities measures is comparable. There are, however, differences when less uncertain outputs were to be linguistically approximated. The similarity measure s_2 reflects the shape of the fuzzy number (and its cardinality) so strongly, that for fuzzy numbers with low uncertainty the linguistic terms $\mathcal{T}_2$ and $\mathcal{T}_4$ will never be used. This is understandable, since for low-uncertain fuzzy numbers closer to the endpoints of the interval $[0,1]$ the asymmetry (and the respective lower cardinality) of T_1 and T_5 respectively outweighs the center-of-gravity information. For low-uncertain outputs s_2 effectively operates only with 3 linguistic terms, which is a considerable reduction of linguistic approximation capabilities. It is also a characteristic that should be considered when choosing the method for linguistic approximation in real-life applications. For example if the linguistic scale used for the approximation contained evaluation terms, the linguistically approximated outputs would be much less rich in variability, thus limiting the discrimination or even decision-support powers of such a model (the effect in this case is one of emphasizing extreme values over the intermediate ones). Note, that even when

$O_i \subseteq T_2$, it will not be linguistically approximated by $\mathcal{T}_2$ (the same holds for T_4 and $\mathcal{T}_4$). This is definitely an undesirable property in many applications.

The remaining distance and similarity measures d_1, d_2 and s_1 perform similarly to each other. None of these measures exhibits the property of extreme-emphasizing. Clearly the influence of shape (cardinality) of the approximated fuzzy number is the highest with s_2 and diminishes through d_2 up to d_1, where it is no longer present. The Bhattacharyya distance d_1 under these circumstances assigns the approximating terms mainly based on the COG, as can also be observed from the strictly vertical boundaries between the colours in the top left plot in Fig. 1.

VI. Conclusion

In the paper we have investigated the role of different distance and similarity measures of fuzzy numbers in the process of linguistic approximation. The results of the presented numerical experiment clearly illustrate, for fuzzy numbers with lower uncertainty, the output of the linguistic approximation depends on the chosen distance/similarity measure. Although this is a well known fact, we have identified several important properties which are relevant for the practical use of linguistic approximation. Firstly if we need to approximate highly uncertain fuzzy numbers, then under our experimental setting there is no apparent difference between d_1, d_2, s_1 and s_2. Second finding identifies the Bhattacharyya distance to be dependent (under the chosen linguistic scale for the linguistic approximation) only on the COG. There is therefore no need to consider the cardinality of the symmetrical triangular fuzzy numbers that are approximated, which can in practical applications mean lower computational costs, since the COG is strong enough information for the linguistic approximation. And the third finding of the numerical experiment is the identification of the extreme-emphasizing effect of the similarity measure s_2 for fuzzy numbers with lower uncertainty.

VII. Acknowledgments

This research was partially supported by the grant IGA PrF 2016 025 of the internal grant agency of the Palacký University, Olomouc.

References

[1] F. Aherne, N. Thacker, and P. Rockett, "The Bhattacharyya Metric as an Absolute Similarity Measure for Frequency Coded Data," *Kybernetika*, vol. 32, no. 4, pp. 363–368, 1998.

[2] D. Dubois and H. Prade, *Fuzzy sets and systems: theory and applications*, Academic Press, 1980.

[3] S. R. Hejazi, A. Doostparast and S. M. Hosseini, "An improved fuzzy risk analysis based on a new similarity measures of generalized fuzzy numbers," *Expert Systems with Applications*, vol. 38, no. 8, pp. 9179–9185, 2011.

[4] A. Kaufman, and M. M. Gupta, *Introduction to Fuzzy Arithmetic*, New York: Van Nostrand Reinhold, 1985.

[5] E. Ruspini, "A New Approach to Clustering," *Inform. Control*, vol. 15, pp. 22–32, 1969.

[6] J. Stoklasa, *Linguistic models for decision support*. Lappeenranta: Lappeenranta University of Technology, 2014.

[7] T. Talášek, J. Stoklasa, M. Collan and P. Luukka, "Ordering of Fuzzy Numbers through Linguistic Approximation Based on Bonissone's Two Step Method," *Proc. 16th IEEE International Symposium on Computational Intelligence and Informatics*, Budapest, Hungary, pp. 285–290, 2015.

[8] S. H. Wei and S. M. Chen, "A new approach for fuzzy risk analysis based on similarity measures of generalized fuzzy numbers," *Expert Systems with Applications*, vol. 36, no. 1, 589–598, 2009.

[9] L. A. Zadeh, "The concept of a linguistic variable and its application to approximate reasoning I, II, III," *Inf. Sci.*, vol. 8, pp. 199–257, pp. 301–357, 1975, vol. 9, pp. 43–80, 1975.

[10] R. Zwick, E. Carlstein and D. V. Budescu, "Measures of similarity among fuzzy concepts: A comparative analysis," *International Journal of Approximate Reasoning*, vol. 1, no. 2, pp. 221–242, 1987.

A K-L divergence based Fuzzy No Reference Image Quality Assessment

Tamim Ahmed
Dept of Computer Science and Engineering
MCKV Institute of Engineering
Howrah,India
yeas1216@gmail.com

Indrajit De
Department of Information Technology
MCKV Institute of Engineering
Howrah,India
Indrajitde2003@yahoo.co.in

Abstract- **Non existence of full proof method of feature extraction of images and lack of well defined relationship between image features and its visual quality leads to erroneous inference in post processing of images. Quality assessment of distorted images without reference to the original image remains an extremely difficult task. Under such circumstances, human reasoning based on subjective information play vital role in assessing quality of image. This paper aims at modeling the quality of distorted images without using any reference to the original image by designing a fuzzy inference system using Kullback Leibler divergence. We determined the probability distribution of mean opinion score(MOS) and also determined the probability distribution of local entropy. Kullback-Leibler divergence is used to determine the distance between two probability distributions. This distance is used in the fuzzy inference system.**

The crisp values of the features and quality of the images are expressed using linguistic variables, which are fuzzified to measure the vagueness in extracted features. Mamdani inference rule has been applied to the FIS (Fuzzy Inference System) to predict the quality of a new distorted image. Quality of test images that incorporated various noises are predicted without reference to the original image producing output comparable with other no reference techniques. Results are validated with the objective and subjective image quality measures.

Key words— **Local entropy, Kullback-Leibler divergence, Fuzzy Inference System, MOS.**

1. INTRODUCTION

Image quality is characteristic of an image that measures the perceived image degradation (typically, compared to an ideal or perfect image). Imaging systems may introduce some amounts of distortion or artifacts in the signal, so the quality assessment is an important problem.

The main objective of image quality metric is to provide an automatic and efficient system to evaluate visual quality. It is imperative that this measure exhibit good correlation with perception by the human visual system(HVS).The ubiquity of transmitted digital visual information in daily and professional life, and the broad range of applications that rely on it, such as personal digital assistants, high-definition televisions, internet video streaming, and video on demand, necessitate the means to evaluate the visual quality of this information. The acquisition, digitization, compression, storage, transmission, and display processes all introduce modifications to the original image. These modifications, also termed distortions or impairments, may or may not be perceptually visible to human viewers. If visible, they exhibit varying levels of annoyance.

Image quality assessment methods can be classified as subjective and objective methods. The first approaches to image quality evaluation are subjective quality testing which is based on observers that evaluate image quality. These tests are time consuming, expensive and have a very strict definition of observational conditions. In Subjective image quality assessment the evaluation of quality by humans is obtained by mean opinion score (MOS) method. It is concerned with how image is perceived by a viewer and gives his or her opinion on a particular image and judge quality of the multimedia content. The MOS is generated by averaging the result of a set of standard, subjective tests. MOS is an indicator of the perceived image quality.

The second approaches are the objective image quality testing based on mathematical calculations and also based on test targets or algorithms. Test-target measurements are tedious and require a controlled laboratory environment. Algorithm metrics can be divided into three groups: full-reference (FR), reduced-reference (RR) and no-reference (NR). This classification is related to the availability of reference images. Over the years, a number of researchers have contributed significant research in the design of full reference image quality assessment algorithms, claiming to have made headway in their respective domains. FR metrics cannot be applied to the computation of image quality captured by digital cameras because pixel-wise reference images are missing. FR metrics are applicable for applications such as filtering or compression where an original image has been processed and the output does not differ in terms of scale, rotation or geometrical distortion. RR metrics provide a tradeoff between NR and FR metrics. An RR metric does not require full access to the reference image; it only needs a set of extracted features. With the aid of RR features, it is possible to avoid problems associated with image content dependencies and multi-dimensional distortion space. .Most images on the internet and in multimedia database are only available in compressed form and hence inaccessibility of the original image, make it difficult to measure the image quality .Therefore ,there is an unquestionable need to develop metrics that closely correlate with human perception without needing the reference image. NR metrics are applicable only when the distortion type is known and the distortion space is low-dimensional. NR metrics are often used to measure a single artifact, such as blockiness, blurriness or motion in video. For example, many NR sharpness metrics interpret graininess or noise as edges or some other image structure. In addition, NR metrics are often highly image content specific. Understanding image content is a difficult task for computational methods. The published methods are still too unreliable for random natural images. We attempt to apply fuzzy inference system to predict the quality of a distorted image. The remainder of the paper comprises of review of existing technologies in section 2. section 3 describes procedure followed in the proposed method and section 4 is about the results and discussions. Conclusion is presented in section 5.

2. REVIEW OF EXISTING TECHNOLOGIES

Full Reference (FR) IQA (Image Quality Assessment)provides a useful and effective way to evaluate quality differences; in many applications the reference image is not available. Although humans can often effortlessly judge the quality of a distorted image in the absence of a reference image, this task has proven to be quite challenging from a computational perspective. *No-reference* (NR) and *reduced-reference* (RR) IQA algorithms attempt to perform IQA with either no information (NR IQA) or only limited information (RR IQA) about the reference image. Here, we briefly survey existing NR - IQA algorithms.

The vast majority of NR IQA algorithms attempt to detect specific types of distortion such as blurring, blocking, ringing, or various forms of noise. For example, algorithms for sharpness/blurriness estimation have been shown to perform well for NR IQA of blurred images.NR IQA algorithms have also been designed specifically for JPEG(Joint Photographic Experts Group) or JPEG2000 compression artifacts. Some NR algorithms have employed combinations of these aforementioned measures and/or other measures .Other NR IQA algorithms have taken a more distortion-agnostic approach.

Numerous algorithms have been developed to estimate the perceived sharpness or blurriness of images. Although the majority of these algorithms were not designed specifically for NR IQA, they have shown success at IQA for blurred images. Modern methods of sharpness/blurriness estimation generally fall into one of four categories: (1) those which operate via edge-appearance models, (2) those which operate in the spatial domain without any assumptions regarding edges, (3) those which operate by using transform-based methods, and (4) hybrid techniques which employ two or more of these methods.

A common technique of sharpness/blurriness estimation involves the use of edge-appearance models. Methods of this type operate under the assumption that the appearance of

edges is affected by blur, and accordingly these methods estimate sharpness/blurriness by extracting various properties of the image edges. For example, Marziliano et al. [1] estimate blurriness based on average edge widths. Ong et al. [2] estimate blurriness based on edge widths in both the edge direction and its gradient direction. Dijk et al. [3] model the widths and amplitudes of lines and edges as Gaussian profiles and then estimate sharpness based on the amplitudes corresponding to the narrowest profiles. Chung et al. [4] estimate sharpness based on a combination of the standard deviation and weighted mean of the edge gradient magnitude profile.Wu et al. [5] estimate blurriness based on the image estimated point spread function. Zhong et al. [6] estimate sharpness based on both edges and information from a saliency map. Ferzli and Karam [7] estimate sharpness based on an HVS-based model which predicts thresholds for just noticeable blur (JNB) the JNB for each edge block is used to estimate the block perceived blur distortions, and the final sharpness estimate is based on a probabilistic combination of these distortions. A related JNB-based method can be found in [8].

In [9], Bovik and Liu presented an NR measure of blockiness which operates in the DCT domain. Blocking artifacts are first located via detection of 2D step functions, and then an HVS-based measurement of blocking impairment is employed.

Park et al. [10] presented an NR measure for blocking artifacts by modeling abrupt changes between adjacent blocks in both the pixel domain and the DCT domain. Similarly, in [11], Chen et al. presented an NR measure of JPEG image quality by using selective gradient and plainness measures followed by a boundary selection process that distinguishes the blocking boundaries from the true edge boundaries.

In [12], Suresh et al. presented a machine-learning based NR approach for JPEG images. Their algorithm operates by estimating the functional relationship between several visual features (such as edge amplitude, edge length, background activity, and background luminance) and subjective scores. The problem of quality assessment is then transformed into a classification problem and solved via machine learning.

Zhou et al. [13] presented an NR algorithm to evaluate JPEG2000 images which employs three steps: (1) dividing the image into blocks, among which textured blocks are employed for quality prediction based on nature-scene statistics;(2) measuring positional similarity via projections of wavelet coefficients between adjacent scales of the same orientation; and (3) using a general regression neural network to estimate quality based on the features from the previous two steps.

Zhang et al. [14] utilized kurtosis in the DCT domain for NR IQAof JPEG2000 images.Three NR quality measures are proposed: (1) frequency-based 1D kurtosis, (2) basis function based 1D kurtosis, and (3) 2D kurtosis. The proposed measures were argued to be advantageous in terms of their parameter-free operation and their computational efficiency (they do not require edge/feature extraction).

In [15], Moorthy and Bovik presented the BIQI algorithm which estimates quality based on statistical features extracted using the 9/7 DWT. The subband coefficients obtained are modeled by a generalized Gaussian distribution,from which two parameters are estimated and used as features.The resulting 18-dimensional feature vectors (3 scales × 3 orientations × 2 parameters) are used to characterize the distortion and estimate quality via the aforementioned two stage classification/regression framework.

In [16], Moorthy and Bovik presented the DIIVINE algorithm, which improves upon BIQI by using a steerable pyramid transform with two scales and six orientations. The features extracted in DIIVINE are based on statistical properties of the subband coefficients. A total of 88 features are extracted and used to estimate quality via the same two stage classification/regression framework.

In [17, 18], Saad et al. presented the BLIINDS-I and BLIINDS-II algorithms which estimate quality based on DCT statistics. BLIINDS-I operates on 17×17 image patches and extracts DCT-based contrast and DCT-based structural features. DCT-based contrast is defined as the average of the ratio of the non-DC DCT coefficient magnitudes in the local patch normalized by the DC coefficient of that patch. The DCT-based structure is defined based on the kurtosis and anisotropy of each DCT patch. BLIINDS-II improves upon BLIINDS-I by employing a generalized statistical model of local DCT coefficients; the model parameters are used as features, which are combined to form the quality estimate.

In [19], Mittal et al. presented the BRISQUE algorithm, a fast NR IQA algorithm which employs statistics measured in the spatial domain. BRISQUE operates on two image scales; for each scale, 18 statistical features are extracted. The 36 features are used to perform distortion identification and quality assessment via the aforementioned two-stage classification/regression framework. Related work on the use of BRISQUE features and discriminatory latent characteristics for NR is found in [20].

In [21] I. De and J. Sill aim to predict the quality of a distorted /decompressed image without using any information of the original image by fuzzy relational classifier. First it clustered the training data using fuzzy clustering method then established a logical relation between the structure of the data and the quality of the image. Quality of a image is predicted in terms of degree membership of the pattern in the given classes applying fuzzy relational operator and a crisp decision is obtained after defuzification of the membership degree.

In [22] I. De and J. Sill presented an algorithm which predict the quality of a image by developing the fuzzy relational qualifier where impreciseness in feature space of training data set is handled using fuzzy clustering method. and the logical relation between the structure of data and soft class labels is established using fuzzy mean opinion score(MOS) weight matrix.

In[23] I. De and J. Sill aim at assessing the quality of distorted/decompressed images without any reference to the original image by developing a fuzzy inference system (FIS). First level the approximation entropies of test images of LIVE database and the features extracted using the five Benchmark images are considered as antecedents while mean opinion score (MOS) based quality of the images are used asconsequent to the proposed FIS. The crisp value of the features and quality of the images are expressed using linguistic variables, which are fuzzified to measure the vagueness in extracted features. Takagi-Sugeno-Kang (TSK) inference rule has been applied to the FIS to predict the quality of a new distorted/decompressed image.

3. PROCEDURE

3.1 Image data set and its subjective MOS

A database of passport size face images are built for training and testing purpose for the proposed algorithm in this paper. This database contains 342 high resolution 8 bits/pixel gray scale distorted (salt &pepper, speckle) images .The database has been divided into two sets to simplify the process of testing and training. Among the two sets, one set is used for training which contains 300 images. Another set is used for testing which contains 42 images.

The proposed algorithm is also validated on Tampere Image Database 2008 (TID2008)[24]. TID2008 contains 17 types of distortions across 1700 distorted images .we take three original images and there corresponding 12 distorted (salt &pepper) images.

Subjects were shown the database; most of them were college students .The subjects were asked to assign each image a quality score between 1 and 5 (1 represents bad and 5 represents good). The 5 scores of each image were averaged to a final Mean Opinion Score (MOS) of the image. Although the purpose of this MOS is to develop **Fuzzy KLD** based **NR** subjective **image quality** assessment **method** .

3.2 Classes of image based on MOS

We partitioned the training set of images into three classes based on MOS(mean opinion score) shown in figure 1 where 1 to 2 represent bad quality , 2.1 to 3.9 represent average quality and 4 to 5represent good quality. GMOS (good mean opinion score), AMOS(average mean opinion score),BMOS(bad mean opinion score) represent three classes of quality for good, average and bad MOS respectively.

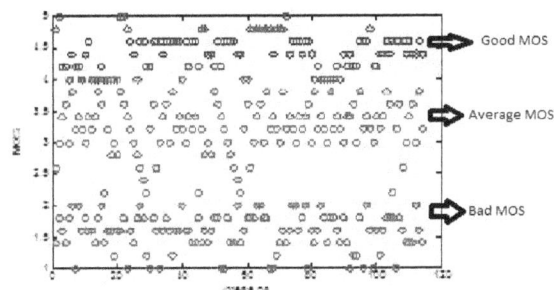

Figure 1: Classes of MOS

3.3 NRIQM System Structure

As shown in Figure 2,

Step 1. We calculated the probability distribution of MOS, GMOS ,AMOS and BMOS..

Step 2. We calculated the probability distribution of local entropy for the distorted images.

Step 3. The probability distributions have been used to find the KLD (Kullback-Leibler distance).

Step 4. Creation of Fuzzy Inference System based on KLD

3.3.1. Probability Distribution

We find out the probability distribution of MOS and three classes of MOS(good, average and bad) GMOS, AMOS and BMOS using the equation (1) as follows.

$$P(x_i) = N(x_i) / n \qquad (1)$$

In the above equation $p(x_i)$ is the probability of each mean opinion score and $n(x_i)$ is the number of occurrence of each score and N is the total number of images. Figures 3 show four probability distributions for GMOS,AMOS, BMOS and MOS respectively. We use Kernel density estimation to estimate the probability density function of a random variable. Kernel density estimation is a fundamental data smoothing problem where inferences about the population are made, based on a finite data sample.

3.3.2. Local Entropy

Local entropy of image means where each output pixel contains the entropy value of the 9-by-9 neighborhood around the corresponding pixel in the input image. For pixels on the borders of the image, local entropy uses symmetric padding. In symmetric padding, the values of padding pixels are a mirror reflection of the border pixels in the image.

We found out the probability distribution of local entropy for three classes of images (good, average and bad) using the equation 1 where $p(x_i)$ is the probability of each pixel entropy and $N(x_i)$ is the no of occurrence of each pixel entropy and N is the total no of pixel. We examined that if the noise increased then the peak of the curve is also increased .From figure 4 and figure 5 we can easily illustrate this.

3.3.3. Relative Entropy (Kullback-Leibler distance)

Let $p(x_i)$ and $q(x_i)$ denote the probability distribution functions of the local entropy of images and mean opinion score, respectively. Kullback-Leibler distance (KLD) between p and q is $D(p\|q)$.

$$D(p\|q) = \sum p \log (p/q)$$

In this paper, we use KLD to quantify the difference between the probability distributions of local entropy of a distorted image and probability distribution of MOS. We get three different KLD range. Based on range we again calculate the KLD between local entropy probability distributions and

probability distribution of each MOS classes (GMOS,AMOS and BMOS) . Here GMOS stands for MOS of good quality images, AMOS stands for MOS of Average quality images and BMOS stands for MOS of bad quality images. The range of KLD obtained for good, average and bad quality class images are 79.25 to 105.39, 46.41 to 61.66 and 93.77 to 169.40 respectively, shown in figure 6. Here all the probabilities is multiplied by 10 .

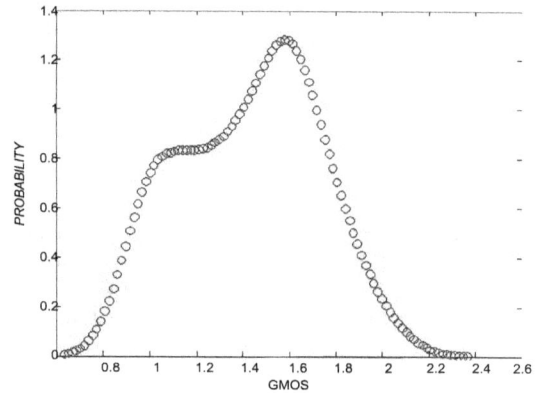

a) Probability distribution of GMOS

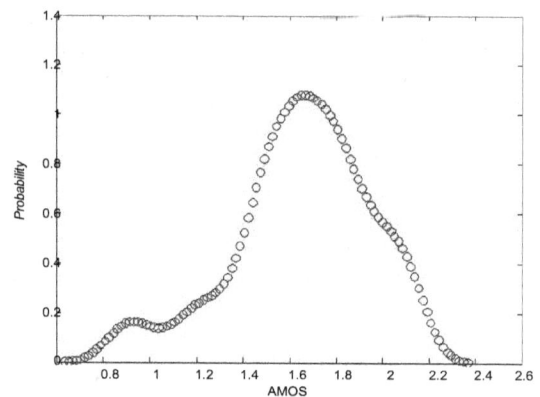

b): Probability distribution of AMOS

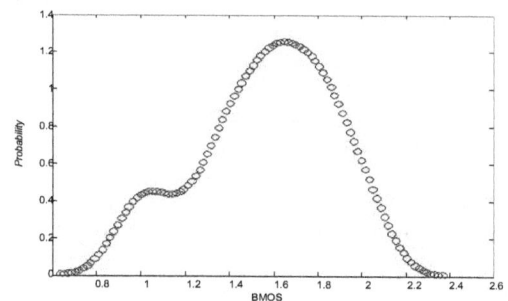

c) Probability distribution of BMOS

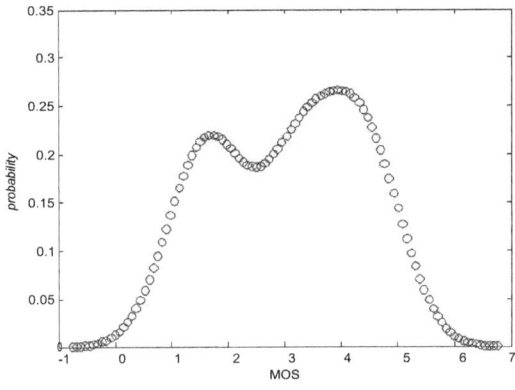

d)Probability distribution of MOS

Figure 3. Probability distribution of GMOS,AMOS,BMOS, MOS

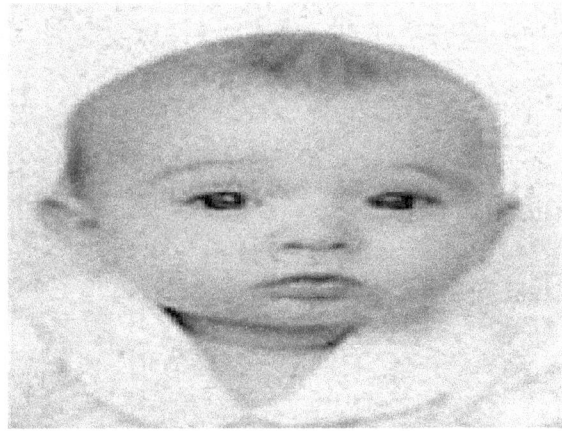

Figure 6: Distribution of KLD

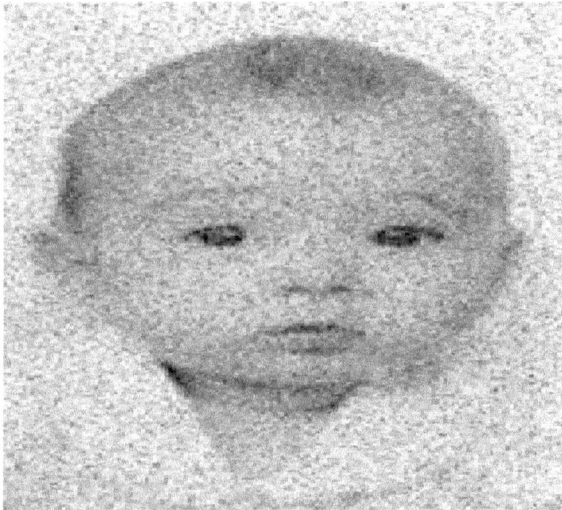

b)A sample image from average class

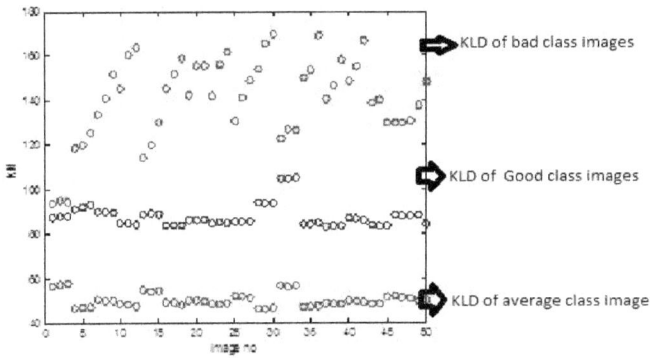

c)A sample image from bad class

Figure 4. Different classes of image

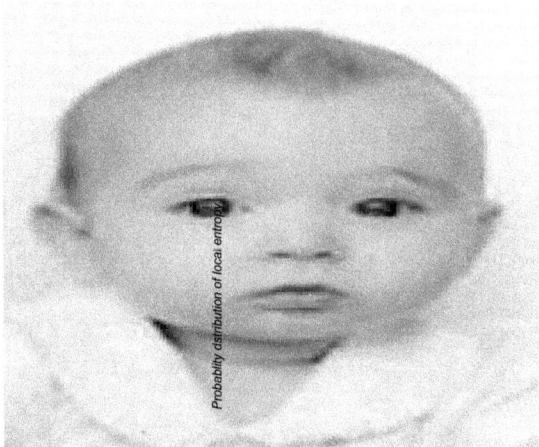

a)A sample image from good class

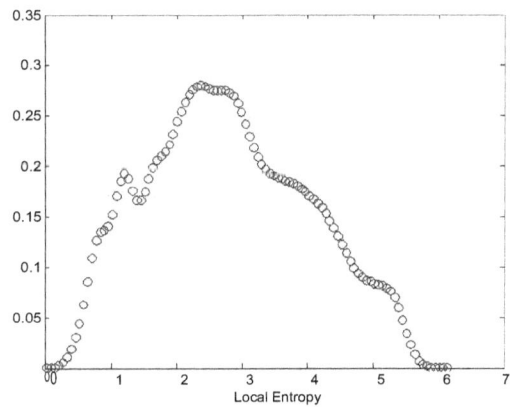

a): Probability distribution of local entropy for good class image

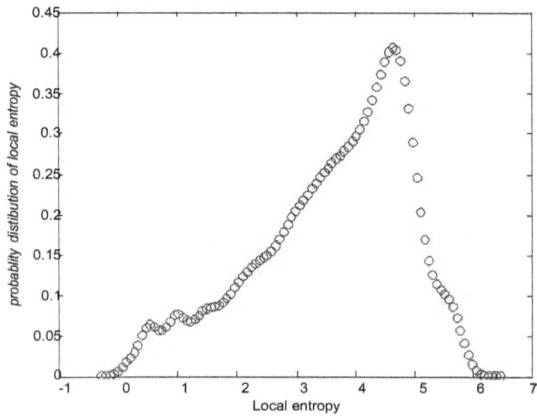

b)Probability distribution of local entropy for average class image

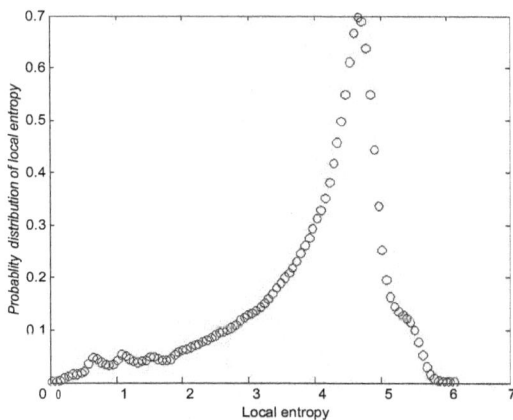

c): Probability distribution of local entropy for bad class image

Figure 5 Probability distribution of local entropy of different classes

The flowchart of the proposed process has been provided. Meanings of some abbreviation are as follows: I/P mf stands for input membership function, Qlt stands for quality, KLD for Kullback-Leibler distance, O/P stands for output. The flowchart is shown in figure 2.

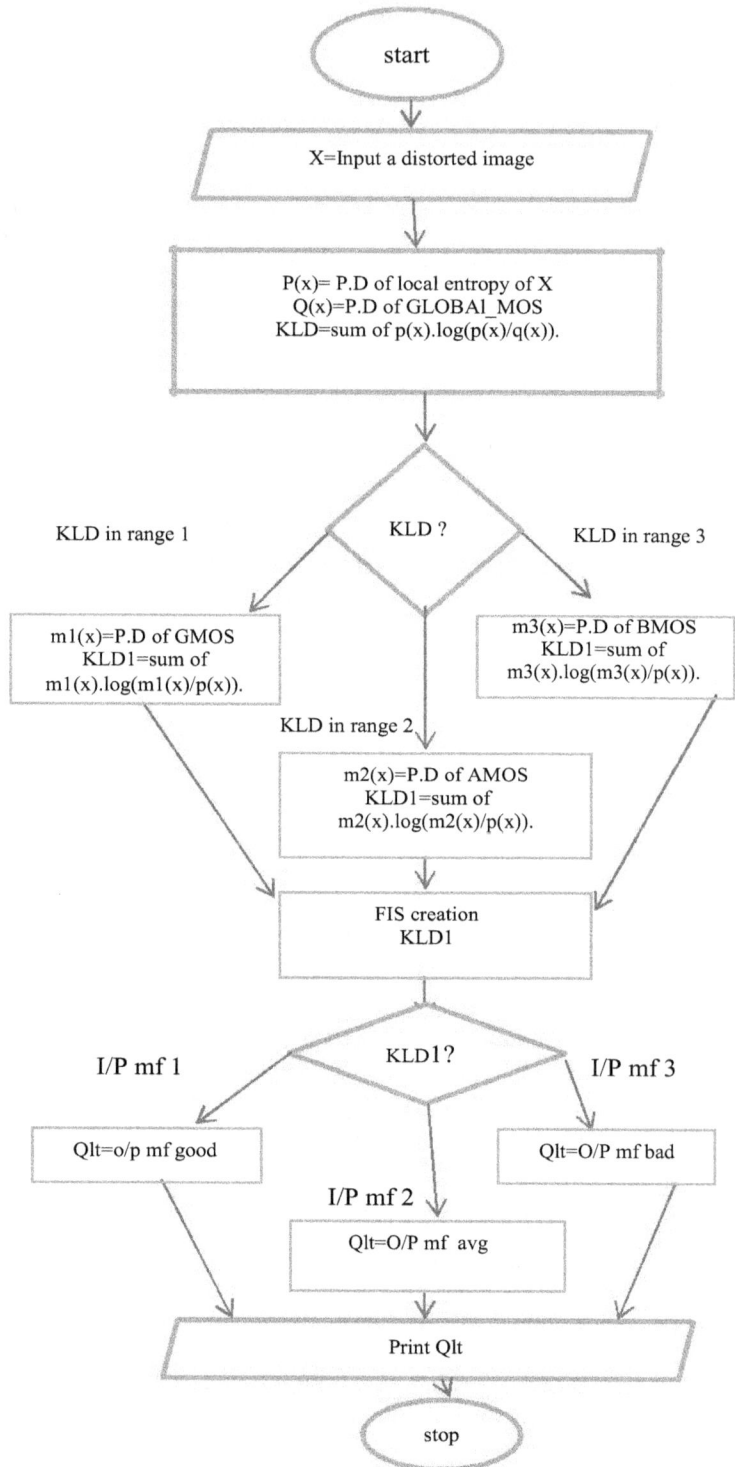

Figure 2.FUZZKLDNRIQM procedure

3.3.4. Fuzzy Inference System for Proposed Model

Mamdani FIS is the most used in the developing fuzzy models. Mamdani architecture used in this paper for estimation of the Quality of an image with one input variable (KLD), one output variable (quality) and three fuzzy rules. The rule base for Mamdani model can be written as FuzzyRule:

If $(x$ is $A_i = M_{1,i})$ {premise}

Then $f_i = M_{o,j}$ {consequent}

Where x represents the input (KLD), linguistic label of input(High, average and bad) and M1i,fi and ith MF of input (x), the output of the jth rule, and the jth output MF. Both input and output MF have their own parameters depending upon the shape of the MF and are called premise, and consequent parameters respectively.

In this paper, the triangle membership function is used to depict membership degrees of the input (KLD1) and output (Qlt) variables.

For the input variable KLD1 we defined three membership functions (high, medium and low) .For high membership function the values of a and c are 79.25, 105.39 respectively from training set of good class images. The value of b is 88.12, calculated by averaging the KLD value of good class. Similarly for medium membership function the value of a,b and c are 46.41,50.34,61.66 respectively. And low membership function the value of a,b,c are 93.37, 140.77 ,169.40 respectively.

For the output variable Qlt we defined three membership functions (good, average and bad) .For the good membership function the parameters value of a is 1,b is 5 and c is 10. Similarly for the average membership function the parameters of a is 10, b is 15 and c is 20 and similarly for the bad membership function the parameter value of a is 20,b is 25 and c is 30.

A. ALGORITHM_FUZZ_KLD_Q

INPUT: IMAGE FOR WHICH QUALITY IS TO BE DETERMINED
OUTPUT: FUZZY QUALITY SCORE OF THE IMAGE

//The variable GLMOS, GMOS, AMOS and BMOS are the arrays denote, respectively, the global mean opinion score, good class mean opinion score, average class mean opinion score and bad class mean opinion score. LE, PLE and PGM are arrays. range1,range2, range3 are defined as follows. $004 \leq$ range1 ≤ 1.18, $0.05 \leq$ range2 ≤ 2.77 and $2.49 \leq$ range3 ≤ 22.60

Begin
Step1: Input a passport size image and store it in the array X Step2: LE ← Local entropy of X.
Step3: PLE ← probability distribution of LE. Step4: PGM ← probability distribution of GLMOS.
Step5: KLD ← Kullback-Leibler distance between PLE and PGM. Step6: If KLD<0
Set KLD ← -1*KLD. Step7:
If KLD in range1
PGMOS ← probability distribution of GMOS.
KLD1 ← Kullback-Leibler distance between PGMOS

and PLE. End if
If KLD in range2
PAMOS ← probability distribution of AMOS. KLD1 ← Kullback-Leibler distance between PAMOS and PLE.

End if
If KLD in range3
PBMOS ← probability distribution of BMOS. KLD1 ← Kullback-Leibler distance between PAMOS and PLE.
End if
Step 8: a) Generate initial model of fuzzy inference system.
b) Add input variable and output variable .
Input variable=KLD1, Range=[46.41 269.40]
Output variable = Quality , Range=[0 30]
c) Set the triangular membership function for input variable KLD1
Membership function1=high, Range=[79.25 88.12 105.39]
Membership function2=medium, Range=[46.41 50.34 61.66]
Membership function3=low, Range=[93.77 140.77 169.40]
d) Set the triangular membership function for output variable Quality
Membership function1=good , Range=[0 5 10]
Membership function2=avg, Range=[10 15 20] Membership function3=bad, Range=[20 25 30]

e) Set the rules between input/outout membership function. R1.
If (kld1 is high) then (quality is good)
R2. If (KLD1 is mediam) then (quality is avg) R3. If (kld1 is low) then (quality is bad)
Step 9: Quality ← fuzzy score based on KLD1. Step 10: Write Quality.
End

4. RESULTS AND DISCUSSIONS

We have tested the proposed No-Reference image quality assessment algorithm in two group of images where group1 contains 20 images from our own database and group2 contains 12 images from Tampere Image Database (TID2008)[24]. We also calculated the no-reference image quality assessment BIQI score and BRISQUE score for group1 and group2 images.

In order to evaluate the performances, two statistical measurements are employed. Linear correlation coefficient (LCC).The Spearman rank-order correlation coefficient (SROCC) evaluates the prediction monotonicity. We found out the Pearson correlation and SROCC value of our proposed algorithm (NRIQM) and MOS and also found the SROCC values of BIQI with MOS and BRISQUESCORE with MOS which are shown in table 1 and table.2. We have plotted the graphs in figure 6 and 7 for MOS and some no-reference IQA for group2 images, group1 images respectively .

Table 1 Pearson correlation between MOS and different NRIQA

Image dataset	Pearson correlation between MOS and BIQI	Pearson correlation between MOS andBRISQUE	Pearson correlation between MOS and NRIQM
Group1	-0.5012	-0.3749	-0.7635
Group2	-0.7434	-.6894	-0.8555

Table 2 SROCC between MOS and different NRIQA

Image dataset	Pearson correlation between MOS and BIQI	Pearson correlation between MOS and BRISQUE	Pearson correlation between MOS and NRIQM
Group1	0.24	0.05	0.26
Group2	0.74	0.42	0.21

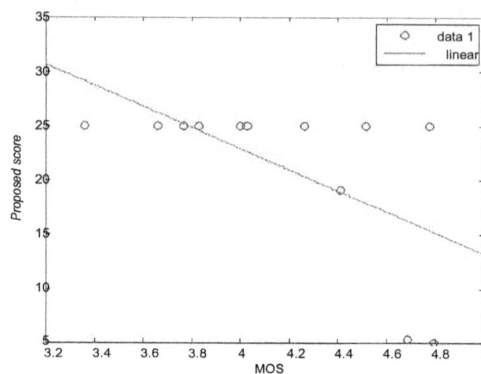

c)Linear relationship between MOS and proposed NRIQA

Figure 6 The linear relationship for group2 images.

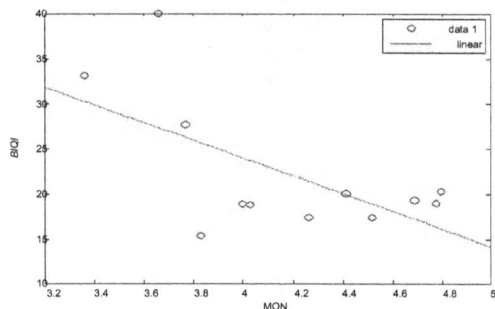

a)Linear relationship between MOS and BIQI

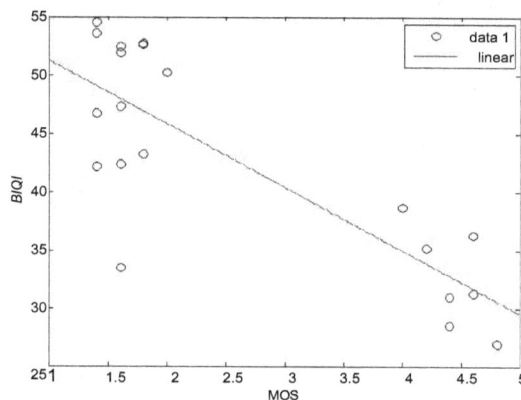

b)Linear relationship between MOS and BRISQUE

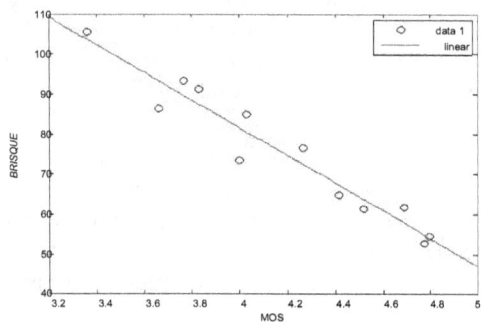

a)Linear relationship between MOS and BIQI

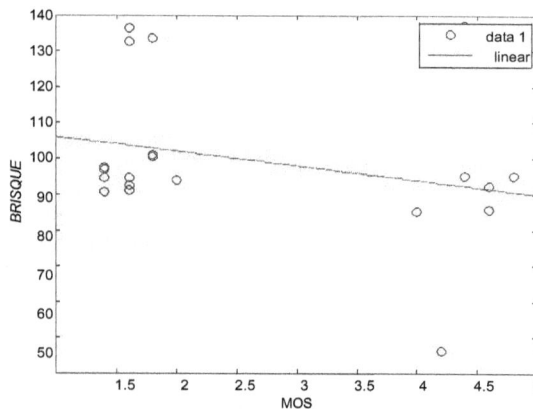

b) Linear relationship between MOS and BRISQUE

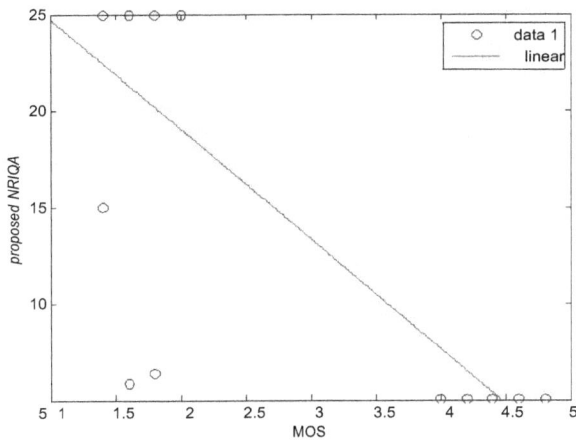

c)Linear relationship between MOS and proposed NRIQM

Figure 7. The linear relationship for group1 images

5. CONCLUSION

We have used type 1 fuzzy inference system to predict the quality of an image based on relative entropy. We have used the concept of local entropy to determine the noise with respect to the mean opinion score given by human .we also calculated the probability distribution of MOS, probability distribution of local entropy. In fuzzy inference system we used triangular membership function for each input variable (relative entropy) and output variable (quality). The peak of input membership functions are the average of each KLD (relative entropy) set .

We have used 300 images for training and 32 images for testing with their corresponding subjective scores. Then, we verified our calibrated results using BIQI, BRISQUE and validated the same using Pearson's correlation coefficient,

SROCC. In future scope of this work eye gaze based real time visually salient dataset is planned to be used and further exploration regarding relationship of such types of data distributions with MOS needs investigation.

BIBLIOGRAPHY

[1]. P. Marziliano, F. Dufaux, S. Winkler, T. Ebrahimi, and G. Sa, "A no-reference perceptual blur metric," in International Conference on Image Processing (ICIP '02), pp. 57–60, September 2002.

[2]. E. Ong, W. Lin, Z. Lu et al., "A no-reference quality metric for measuring image blur," in Proceedings of the 7th International Symposium on Signal Processing and Its Applications, vol. 1, pp. 469–472, 2003.

[3]. J. Dijk, M. Van Ginkel, R. J. Van Asselt, L. J. Van Vliet, and P. W. Verbeek, "A new sharpness measure based on Gaussian lines and edges," CAIP, vol. 2756, pp. 149–156, 2003.

[4] .Y. C. Chung, J. M. Wang, R. R. Bailey, S. W. Chen, and S. L. Chang, "A non-parametric blur measure based on edge analysis for image processing applications," in Proceedings of the IEEE Conference on Cybernetics and Intelligent Systems, pp. 356–360, December 2004.

[5].S. Wu, W. Lin, L. Jian, W. Xiong, and L. Chen, "An objective out-of-focus blur measurement," in Proceedings of the 5th International Conference on Information, Communications and Signal Processing, pp. 334–338, December 2005.

[6]. S. H. Zhong, Y. Liu, Y. Liu, and F. L. Chung, "A semantic no-reference image sharpness metric based on top-down and bottom-up saliency map modeling," in Proceedings of the 17th IEEE International Conference on Image Processing (ICIP '10), pp. 1553–1556, September 2010.

[7]. R. Ferzli and L. J. Karam, "A no-reference objective image sharpness metric based on the notion of Just Noticeable Blur (JNB)," IEEE Transactions on Image Processing, vol. 18, no. 4, pp. 717–728, 2009.

[8]. N. D. Narvekar and L. J. Karam, "A no-reference perceptual image sharpness metric based on a cumulative probability of blur detection," in Proceedings of the International Workshop on Quality of Multimedia Experience (QoMEx '09), pp. 87–91, July 2009.

[10]. C. Perra, F. Massidda, and D. D. Giusto, "Image blockiness evaluation based on sobel operator," in Proceedings of the IEEE International Conference on Image Processing (ICIP '05), pp. 389–392, Genova, Italy, September 2005.

[11]. H. Zhang, Y. Zhou, and X. Tian, "Weighted sobel operator-based no-reference blockiness metric," in Proceedings of the Pacific-Asia Workshop on Computational Intelligence and Industrial Application (PACIIA '08), pp. 1002–1006, Huhan, China, December 2008.

[12]. C. S. Park, J. H. Kim, and S. J. Ko, "Fast blind measurement of blocking artifacts in both pixel and DCT domains," Journal of Mathematical Imaging and Vision, vol. 28, no. 3, pp. 279–284, 2007.

[13]. "No reference image quality assessment for JPEG2000 based on spatial features," Signal Processing: Image Communication, vol. 23, no. 4, pp. 257–268, 2008.

[14]. H. R. Sheikh, A. C. Bovik, and L. Cormack, "No-reference quality assessment using natural scene statistics: JPEG2000," IEEE Transactions on Image Processing, vol. 14, no. 11, pp. 1918–1927,2005

[15]. "No-reference image quality assessment using visual codebook," IEEE Transaction on Image Processing, vol. 21, no. 7, pp. 3129–3138, 2012.

[16]. A. C. Bovik, "Automatic prediction of perceptual image and video quality," Proceedings of the IEEE. In press.

[17]. M. Saad, A. Bovik, and C. Charrier, "A DCT statistics-based blind image quality index," IEEE Signal Processing Letters, vol. 17, no. 6, pp. 583–586, 2010.

[18]. M. A. Saad and A. C. Bovik, "Blind image quality assessment: a natural scene statistics approach in the DCT domain," IEEE Transactions on Image Processing, vol. 21, pp. 3339–3352, 2012.

[19]. A. Mittal, A. K. Moorthy, and A. C. Bovik, "No-reference image quality assessment in the spatial domain," IEEE Transactions on Image Processing, vol. 21, no. 12, pp. 4695–4708, 2012.

[20] .A. Mittal, G. S. Muralidhar, J. Ghosh, and A. C. Bovik, "Blind image quality assessment without human training using latent quality factors," IEEE Signal Processing Letters, vol. 19, no. 2, pp. 75–78, 2012.

[21]. Indrajit De , Jaya Sill "No Reference Image Quality Assessment Using Fuzzy Relational Classifier" AICI 2011 ,Part 1 LNAI 7002,pp.551-558,2011

[22]. Indrajit De ,Jaya Sill "No Reference Image Quality Assessment by designing Fuzzy Relational classifier Using MOS Weight Matrix" AICI 2011,LNBI 6840,pp.369 2012

[23] Indrajit De , Jaya Sill "Wavelet entropy based no-reference quality prediction of distorted/decompressed images" Computer Engineering and Technology (ICCET), 2010 2nd International Conference on (Volume:3)

[24] N. Ponomarenko, V. Lukin, A. Zelensky, K. Egiazarian, M. Carli, F. Battisti, "TID2008 - A Database for Evaluation of Full-Reference Visual Quality Assessment Metrics", Advances of Modern Radioelectronics, Vol. 10, pp. 30-45, 2009.

Profitability Analysis of Small and Medium Enterprises in Romania Using Neural and Econometric Tools

Dumitru-Iulian NASTAC
Department of Electronics, Telecommunications and Information Technology
POLITEHNICA University of Bucharest
Bucharest, Romania
nastac@ieee.org

Irina-Maria DRAGAN, and Alexandru ISAIC-MANIU
Department of Statistics and Econometrics
Bucharest University of Economic Studies
Bucharest, Romania

Abstract—The SMEs sector in Romania, still ongoing consolidation, is particularly vulnerable to the specific factors of the market economy fluctuations, an economy not yet fully functional as a consequence of the transition from planned communist type economy, which had operated nearly half a century. The purpose of the performed analysis is to identify some potential factors for influencing the companies increasing performance, in order to provide several foundation elements for the governmental policies and strategies in the field. The study uses exhaustive balance sheet information, not obtained by sampling, and refers to the strongest SMEs segment, the medium-sized enterprises (n=7,902 units of a total of N=572,800 SMEs with approved financial documents), these being likely to be closer to the functioning company's mechanism. The balance sheet data processing, for the medium-sized enterprises, was performed using an artificial neural networks (ANNs) model which was validated by comparing its outcome with other results provided by multiple statistical regression models. The main goal was to establish a flexible ANN classifier, for these medium-size enterprises, which can be further adapted to eventually nonstationary changes. A three-class ANN-model was found as a good option after analyzing the data. These first results are encouraging for further developments in order to obtain an adaptive system.

Keywords—SMEs, neural networks, classification, econometric models, strategic measures, validation tests

I. THE ROMANIAN SMEs SECTOR IN THE EUROPEAN CONTEXT

In Romania, as well as in many other ex-communist countries, the transition to market economy essentially meant the evolution of several components: transition from state to mixed proprietorship, transition from planned mechanisms to supply-demand mechanisms, from strict price-control to a price given by demand and supply, a change in how wealth and newly created value is being distributed, a change in how the society is divided into classes, transition from dictatorship to democracy. Of all these processes, perhaps the most important one in matters of economic consequences was the change in property structure, realized in two ways: the transfer of property and the stimulus of free initiative, leading to the appearance of new enterprises. If the transfer of property, i.e. the privatization of state enterprises, is a process which approaches an end, private initiative has plenty of room left for development, if we were to compare to consolidated market economies. The use of the entire potential of SMEs is an essential part of the EU strategy, materialized in 2008 through the Small Business Act for Europe [1].

The Romanian SME sector has been strongly affected by the global recession from 2008 to 2010 and has managed to slightly recover afterwards, only to be slowed in 2012 [2] [3]. For 2014 and 2015 we expect an increase which should return the sector to prior-to-crisis levels regarding number of firms and employees [4].

The *rate of new enterprise creation*, an indicator computed for EU countries starting with 1995, and for Romania starting 2000 (the base being the stock of firms in 1995 = 100%), has constantly increased until 2008, when it reached 43.3%, while in the last two years it reached 35%. The *rate of firm mortality*, computed as the report of closed firms and their stock, has high values in Lithuania (around 30%) and low in Cyprus (around 3%), while in Romania it's between 10-12%. The *firms' survival rate* has reduced values in Lithuania (around 55-60%), Slovakia (50-55%), and very high values in Sweden (90%). Romania is placed close to the upper part of the ranking, reaching the rate of 80% after 4 years since creation, with some differentiations depending on the industry [5] [6].

The volume of the 578800 SMEs active in 2013 represents a major weight, of 99.7%, in the total number of active firms in Romania, with an increasingly contribution to the creation of workplaces, quantified at around 60% of the total number of employees. In 2012, the number of SMEs decreased by 7.0% versus 2011 [7].

Regarding the role played in the national economy, expressed through the mentioned percentages, Romanian SMEs are generally placed at the same level as other former socialist countries in the EU [8].

The number of SMEs per number of inhabitants places Romania among the countries with low density, with 36.8

63

SMEs/1000 inhabitants, which means 60% of the European average. Moreover, there is a lack of balance among the eight regions of the country, which emphasizes development differences across regions: the region Bucharest-Ilfov has 24.5% of the firms, while the South-Western region has 7.4%; similarly, entrepreneurial density is: 62.1% SMEs per 1000 inhabitants in Bucharest-Ilfov and 15.6% in the North-East region [9].

The structure by SMEs size class is: micro-enterprises 90.6%, small enterprises 7.68%, and medium-sized enterprises 1.36%. The number of the 2.451.183 employees is equally distributed to the three classes: 32.5% in micro-enterprises, 35.3% in small enterprises and 32.2% in medium-sized enterprises. The average number of employees per firm is 4.28 at the level of the entire sector, with some differentiations: 1.53% in micro-enterprises, 19.63% in small enterprises and 101.35% in medium-sized enterprises.

Gradually, SMEs have diversified their activity profile, presently 88 of the NACE activities being active. With over 5% weights of the total number there are: retail trade (21.5%), wholesale trade (10.5%), land transports and via pipelines transports (6.0%) and building constructions (5.5%).Most of the firms specialized in high-end technological manufacturing without elements specific to knowledge economy. Compared to the EU average, Romania has much less medium-tech industries – such as the manufacturing of chemical substances, electrical equipment, cars and transportation equipment. For instance, the average of the commercial SMEs weight in total EU is about 30%.

The consolidation of this economic sector becomes a priority of governmental policies as the economic crisis demonstrated its importance in an increasingly competitive unique European Union market. At the same time, national policies must be aligned at European level in the SMEs and entrepreneurship support frame considering that Romanian SMEs compete with European companies in a common market – the EU internal market – with equal standard requirements.

The goal of our research was to find a useful model able to capture the complex relationships among various specific parameters that can describe or influence companies' profit in order to finally classify them. In order to identify some supporting elements for the establishment of some strategies to strengthen the SMEs sector and increase their profitability, the analysis was focused on the strongest segment of SMEs, the medium-sized companies, represented by a number of $N = 7,902$ lines of data, which were firstly used to build and evaluate the neural tool and then for the regression models, as alternative validation process of the results obtained in the first phase.

The structure of the paper is as follows. Section 2 presents the problems that concerned the literature review in connection with our aim. A description of the data, together with the methodology involved in this work and the main features of our experimental results are given in the next section, where we also discuss a comparison with a classic approach. Conclusions are formulated in the last part of our paper.

II. LITERATURE REVIEW

The importance of SMEs to the economy is given by various points of view, including the contribution to national production, the use of internal resources, the increase of labor force occupancy, the ease of workforce migration at local and European levels, the increase of national economy competitiveness, and the better use of human capital. This context quite explains the increasing interest of not only decision makers in the economy but also of the political and academic circles.

Thus, a special attention is given to key-factors which determine the increase of the number of SMEs, using the relationship between economic and financial growth, the way in which SMEs overcame the crisis, econometric modeling of firms' growth and the influencing factors. Exploring the main decisive factors for the growth of SMEs in CEE countries was done by Mateev and Anastasov [10] for which they used a data panel of 560 rapidly growing companies from six emerging economies. Important factors, with leverage effect, are liquidity rates, future growth opportunities, labor productivity. An extended analysis on SME sector was made by Ayyagari, Beck, Demirguc-Kunt [11] on a number of 76 countries, and it highlights the relation between the size of the SMEs sector and the business environment. The paper presents coherent and comparable information regarding the contribution of the SMEs to the total number of work places in the manufacturing industries and in GDP, including the informal economy activity. An analysis of the Bulgarian SMEs is published by the Bulgarian Small and Medium Enterprises Promotion Agency [12], the Bulgarian profile authority.

Jennings et al. [13] investigates the connection between the performance of the firms and entrepreneurial spirit on the basis of the data from some companies operating in the area of the industry of services based on the knowledge, by deepen the study using quantitative methods, which allows the formulation of the relevant conclusions on the subject matter of the investigation. Zhou and de Wit [14], having an integrated analysis of the matter, established that the firms' growth determining factors are classified in three categories: individual, organizational and environmental. Lejárraga et al. [15] explore issues regarding the business internationalization of manufacturing SMEs and their various related services. Based on the experimental results, is highlight the link between the firm size and the business performance, less obvious in the export outcome. Also, are emphasizing the situations in manufacturing firms and those in services. An empirical study of 523 small and medium-sized Dutch firms has identified the factors which determine firm growth, measured as an increase in workforce occupancy. The organizational factors have the greatest influence on firm growth, as well as financial capital availabilities. Heshmati [16], using data for Swedish industrial SMEs with up to 100 employees and high weights versus the activity of the region, a major factor which seems to influence firm growth is a positive indebtedness ratio which leads to sales growth. Also, the impact of various regional development programs and policies for the period 1993-1998 was analyzed. Measuring the rate of growth was based on the number of employees, sales level and assets volume and the control factors were: capital structure, performance, human capital as well as local conditions in the labor market. The data was used as entry information in econometric models and the conclusions are interesting given possibilities of extension and interpretation.

An investigation of the effects of the Agricultural Guidance and marketing of entrepreneurial information on the performance of small and medium-sized enterprises was performed by Keh, Nguyen, and Ng [17], through the construction, testing and validation of a model of the causal link using the data obtained from the entrepreneurs from Singapore. The conclusion of the authors indicates that the orientation of the venture an important role play using the information of marketing with positive effects on the performance of the firms. An assessment of the impact of SME sector growth is given by Subhan, Mehmood, and Sattar [18], mainly focused on the role of innovation and the effects on the Pakistani economic development. For measuring innovation, the authors propose the level of C&D spending, number of patents, number of publications, technological intensity, and high-technology exports. Other variables included in this study are the weight of exports in GDP, the increase of SMEs, GDP growth, industry weight in GDP, the level of workforce occupancy, the consumption price index, the exports and imports volume and the exchange rate. The valid econometric model was a linear regression for the analysis period 1980-2012, and the results indicate to a positive correlation between innovation and the increase of SMEs.

The issue of firm growth is also approached by Evans [19] who examined the effects of a firm's dimensions and age, using data on 100 US manufacturing firms. The author reached the conclusion that a firm's age and size are inversely proportional to its growth chances. The uncertainty raised by predictions on macro-economy evolution was identified by Henzel and Rengel [20] as being caused by the fluctuations of business cycles, oil and raw material prices.

Using the information from a sample of 144 small and middle-sized companies in China and the involvement of the corporate social responsibility(CSR), in [21] it was considered the environmental performance in conjunction with the size of your company, as well as the differentiation of the involvement in the programs of social responsibility in the light of the economic power of them.

A relatively new problem for SME management is the issue of networks, tackled by Terziovski [22] who, having went over the experience of large corporations, described a model for network assembly. Hussey and Eagan [23] propose econometric methods for formulating environment policies for SME activity, using structural equations in order to develop environmental performance models for SMEs, at the same time supplying an application based on the case of some plastic manufacturers. The elaboration of CSR strategies for SMEs is developed by Perri, Russo, and Tencati [24], as this behavioral component is usually attributed to large enterprises. The authors make a comparison between large and small or medium-sized firms using the results of an investigation made on a sample of 3680 Italian firms. An interesting application was performed [25] in connection with the economic crime rate, with general offenses and their effects on economic

growth rates. The study is based on the data of over 12,000 companies, from 27 developing countries, and the findings obtained reveal a negative correlation between the economic crime and economic growth rate, more evident being the impact in the small companies category. An analysis of the financial crisis has strongly affected large companies (LMES) as well, generating a space for SME extension, internally as well as at international level, shows the conclusions of Tayebie et al. [26], based on a study on Asian economies which used panel data sampling for a gravitational econometric model (the same economies were analyzed by Elliot and Ikemoto [27]). In the same field, Terdpaopong [28] proposes to determine whether a statistic model can in fact identify the crisis of a firm's debts. A sample of 159 SMEs was used, including some firms with financial difficulties, as well as others which don't have this problem. Using a logistic regression model, validated by specific testing, is used to determine the probable chance of survival or failure using predictive models. The authors try to use a predictive model to anticipate firms with financial difficulties and to support correction decisions regarding the financial behavior. An interesting analysis on the differences in the economic performance of small and medium-sized enterprises according to the genus self was performed by Watson and Robinson [29]. The authors have taken into account many variables (compared with studies of this type which is limited to indicators as well as profit and/or increase of turnover) among which the age of the undertaking and of the developer, the size of the company, the risk business type. The authors arrive at the most interesting conclusions referring to the fact that there is no significant differences of results between undertakings after the genus entrepreneurs, if it corrects the data according to the risk, which practically breaking the myth concerning the superiority of self-made business performance. The question of the different performance classification has been tackled in the literature during the last decades. Some useful taxonomies of neural classification models are presented in [30] and [31]. Before that, a series of statistical techniques have been developed such as: univariate statistics [32], multivariate analysis [33], linear discriminant analysis introduced by Fisher [34], multivariate discriminant analysis [35], and many others. Another way for solving the classification problems it was the use of the induction techniques, such as recursive partitioning algorithm [36], CART [37], and ID3 [38]. The self-organizing map (SOM) model was introduced by Kohonen in early 80's [39] and is an unsupervised learning technique used in classification, which creates a two-dimensional topological map from n-dimensional input data. Many other authors developed applications of artificial intelligence models, like neural networks [40] [41], genetic algorithms [42], machine learning [43] [44], etc. Several hybrid solutions were also proposed [45] [46] [47] in order to solve classification problems.

III. WORKING METHODOLOGY AND MAIN RESULTS

In this section we present an extension of the preliminary results published in [48]. All data involved in this study come from the Romanian National Statistics Institute for synthetic indicators and for SMEs demography indicators, while data from the Romanian trade Registry were used for balance sheets. We were focused to further analyze the category of medium-sized enterprises, eliminating database of micro and small enterprises, considering that for these firms there is a greater sensitivity of results and performance compared to the volume of assets, the personnel expenses, financial expenses, the taxes volume and other balance sheet indicators, that condition the size of the final results.

Following validation of balance sheets, the volume of medium-sized firms in the analysis was 7902. We selected as indicators related to these companies: the company location – the county, ownership status, regional territorial area, type of economic activity according to international classification (NACE), and a set of balance sheet indicators.

Therefore, the variables under consideration were: counties (jj); property type (pp); NACE codes (caen); total fixed assets (AIT); total current assets (ACT); debts: amounts due in a period up to a year (DSA); total assets minus current debts (TAMDC); debts: amounts due in a period larger than one year (DPA); total owner's equity (CPT); total equity (CT); personnel expenses (CHP); total operating expenses (CHE) total financial expenses CHF) ; total expenses (CHF) corporate tax (IMP) ;average number of employees (NMS) development region (reg).

The processing and interpretation of results was made using neural networks and regression econometric models for establishing the conclusions.

Firstly, a special classifier model based on artificial neural network (ANN) architecture was employed in order to capture distinctive aspects of the dataset. Basically, this is a classification tool, which may include (even through a future extension) many important parameters. Frequently, this issue is analyzed in literature as a binary classification, also identified as the two-group discriminant problem. This is, of course, the simplest case of the grouping problem. In this case everything is viewed in black and white. A model which implements a binary classifier would just show a good or bad performance for an economic player, without giving detailed information about its real problems. Obviously, better information can be obtained out from the classification model if we can divide its output into more than two performance classes. Then, it would easier to analyze the companies according with their positions in these classes. Therefore, we have considered three categories (poor/ medium/ high) as a result of the splitting in intervals which denote economic performance according to profitability rate. This parameter has a range of values starting from zero to almost two hundred. In order to split the companies in the previously mentioned categories we supplementary selected two numbers as delimiters B_{12} and B_{23}, where the numerical indices (each pair of digits as subscript of B) consecutively indicates two neighbor classes for a specific delimiter. For example we can consider $B_{12}=0.5$ and $B_{23}=5$, but we can vary these values or, if necessary, expand the number of classes. To be more explicit, having only three classes as in previous mentioned example, those firms whose parameter profitability rate varies from zero to $B_{12}=0.5$ (inclusive) will fall in the poor class. Next, we can classify as medium-performance all companies which indicate values strictly greater than $B_{12}=0.5$ till maximum $B_{23}=5$. And finally the last group (with best

performances) includes everyone that has the mentioned parameter strictly greater than $B_{23}=5$.

One of the major problems was how to establish the borders between classes. We have to start with a first attempt and then there may be interesting to study what it is happening when changing the borders between these successive classes. Different changes of the borders will be taken into consideration. Supplementary, one can further extend the number of classes if it is intended to capture detailed aspects of firm performances.

Note that a big challenge in this attempt is to properly classify those companies which are very close to the borders. Important is how to capture a proper correlation between input parameters and the output of the classification model in order to minimize wrong classification especially in the vicinity of the borders.

The ANN classifier model includes several feature especially designed in order to obtain maximum performances by capturing some properties of the dataset. Figure 1 schematically shows the model used in our attempt. There is depicted a neural network during its training process.

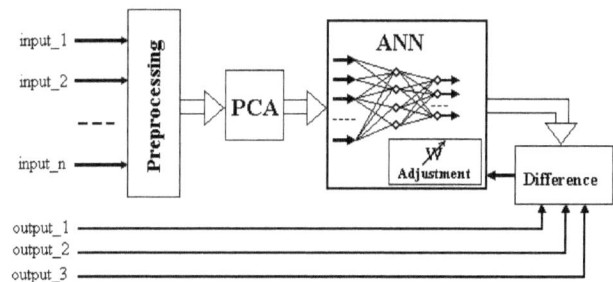

Fig. 1. The ANN classification model during the learning process

In this model, each output corresponds to one specific class. The inputs are preprocessed by using normalization and then a principal component analysis bloc (PCA) [49] [50] is employed in order to reduce the dimension of input space. During the learning process, the neural network has to extract and use the subtle correlations that exist between the model's outputs and the whole inputs. For a specific configuration, when a class is selected, then the value of the corresponding output has the value of one, while the rest of outputs are zeros.

Frankly speaking, it is difficult to find which ANN training algorithm will provide the best result for a specific problem. A good choice depends on many parameters of the ANN involved, the data set, the error goal, and whether the network is being used for pattern recognition (classification) or function approximation. It seems that for large datasets the numerical optimization techniques present numerous advantages. Analyzing the algorithms that fall into this class, we found that the Scaled Conjugate Gradient (SCG) algorithm [51] performs very well over a wide variety of problems, which also include the one presented in this paper. Even if the SCG algorithm is not the fastest one (like Levenberg-Marquardt [52] in some situations) the great advantage is that it works very efficiently for networks with a large number of weights. The SCG is an excellent compromise since it does not require large computational memory, and yet it still has a good convergence and is very robust. Moreover, we always apply the early stopping method (validation stop) during the training process, in order to avoid the overfitting phenomenon. And it is well known that for early stopping, one must be careful not to use an algorithm that converges too rapidly, in order to avoid blocking into a local minimum. Therefore, the SCG is well suited for validation stop method.

An iterative process was used in order to find the proper architecture of the ANN model. Here, a series of imbricate loops were employed to get the number of neurons on each layers. We finally used feedforward ANNs with two hidden layers in order to achieve a good classification function, based on preliminary research [41] [48], where it has been obtained better results in case of two hidden layers than in case of a single hidden layer, however maintaining a somehow similar ratio (approx. 10/1, or greater for present data) between the number of the training samples and the total number of the weights. Consequently, we used only two imbricated loops for searching for the proper number of neurons for each hidden layer (N_{h1} and N_{h2}). Each of the training sessions started with the weights initialized to small, uniformly distributed values. We tested several ANN architectures, with N_{h1} and N_{h2} taking values somehow in the vicinity of the geometric mean of the neighboring layers, and trying to observe if the following rules can be respected:

$$\sqrt{N_i \cdot N_{h2}} + q \geq N_{h1} \geq \sqrt{N_i \cdot N_{h2}} - q \qquad (1)$$

$$\sqrt{N_{h1} \cdot N_o} + q \geq N_{h2} \geq \sqrt{N_{h1} \cdot N_o} - q \qquad (2)$$

Here, N_i is the number of inputs after PCA block, N_o is the number of outputs, and q is a small positive integer (ex. q=4). As a consequence, we independently vary the number of the neurons, from each hidden layer (between 4 and 10 for the first hidden layer, and from 2 to 8 for the second hidden layer), in order to cover a proper number of cases, but maintaining a decent time and reasonable computational effort. Each of the ANN architecture was tested five times, with random initial settings of the weights and different training/validation sets. We chose the best model with respect to the smallest error between the desired and the simulated outputs. This error (E_{tot}) was calculated for both training and validation data sets. By using this way, we selected the optimum network from a total of 7×7×5=245 different instances. Furthermore, we introduced the supplementary condition:

$$E_{val} \leq \frac{6}{5} \cdot E_{tr} \qquad (3)$$

in order to avoid a large difference between the error of the training set (E_{tr}) and the error of the validation set (E_{val}). We simply have to remove those combinations which provide a value of E_{val} that is with 20% greater than E_{tr}, during the validation stop. This way, the overfitting phenomenon on the test set will be considerably reduced. In our approach the validation set acts at the same time as a kind of test set (having in mind that theirs associated errors are pretty similar), even though there is a totally separate test set.

In order to obtain relevant results, we have randomly chosen, from the initial set of data (7902 samples), the following three subsets with their approximate percentages: ~70% for the effective training process, ~10% for validation, and ~20% for test.

TABLE I. RANDOM SPLITTING OF THE DATA FOR EACH SEPARATE TRIAL

	B_{12}	B_{23}	N_{tr}			N_{val}			N_{test}		
			Class 1	Class 2	Class 3	Class 1	Class 2	Class 3	Class 1	Class 2	Class 3
Trial 1	0.5	5	1434	1850	2338	182	247	331	413	461	646
Trial 1 (second iteration)	0.5	5	1461	1837	2357	174	261	314	394	460	644
Trial 1 (third iteration)	0.5	5	1477	1804	2359	204	253	297	348	501	659
Trial 2	1.5	7	1852	1948	1852	245	264	241	495	514	491
Trial 2 (second iteration)	1.5	7	1843	1928	1872	250	265	238	499	533	474
Trial 2 (third iteration)	1.5	7	1838	1916	1835	268	270	233	486	540	516
Trial 3	1.5	8	1827	2172	1632	242	289	226	523	555	436
Trial 3 (second iteration)	1.5	8	1828	2166	1634	264	274	220	500	576	440
Trial 3 (third iteration)	1.5	8	1886	2151	1630	227	293	225	479	572	439
Trial 4	1.55	6.85	1881	1866	1896	259	254	240	493	514	499
Trial 4 (second iteration)	1.55	6.85	1872	1860	1905	252	262	241	509	512	489
Trial 4 (third iteration)	1.55	6.85	1887	1908	1848	239	240	274	507	486	513
Trial 5	2	8	2085	1896	1626	301	246	218	536	544	450
Trial 5 (second iteration)	2	8	2077	1954	1606	277	255	223	568	477	465
Trial 5 (third iteration)	2	8	2120	1904	1631	270	246	233	532	536	430
Trial 6	1	10	1653	2729	1267	203	365	183	455	688	359
Trial 6 (second iteration)	1	10	1633	2666	1311	227	374	163	451	742	335
Trial 6 (third iteration)	1	10	1659	2735	1300	204	371	161	448	676	348

Under these circumstances it is quite important to have enough data for every kind of processes (training, validation and test). Furthermore, when we have to split the data into classes is also significant to have a somehow balanced distribution. Otherwise a class with a greater amount of data will influence the result (by capturing the false matching from the less representative classes) and during the test process this will have a negative impact over the results. Therefore, in order to keep this splitting under observation, we established successive trials with specific predefined borders, each of these trials being repeated three times to see if the randomization is effective. These distributions for 6 trials (totalizing 18 sub-trials) are presented in Table I.

There is a reasonable distribution of the data in all trials. According with the Table I it seems that both second and four trials show almost equal data partitions for each class. It is worth to mention that the selection of the borders between successive classes was intuitively suggested by specialists in economics and we wanted to see the result of this approach. Then we have to construct an ANN structure for each of these trials.

TABLE II. THE RESULTS OF DIFFERENT APPROACHES IN WHICH WE VARY THE BORDERS (B_{12} AND B_{23}) BETWEEN CLASSES

	B_{12}	B_{23}	N_{tr}	N_{val}	N_{test}	$P_{Class\ 1}$	$P_{Class\ 2}$	$P_{Class\ 3}$	P_{Test}	N_{h1}	N_{h2}
Trial 1	0.5	5	5622	760	1520	0.2833	0.6356	0.8375	0.6257	8	7
Trial 1 (second iteration)	0.5	5	5655	749	1498	0.0127	0.7217	0.8509	0.5907	6	5
Trial 1 (third iteration)	0.5	5	5640	754	1508	0.2155	0.6766	0.8027	0.6253	8	6
Trial 2	1.5	7	5652	750	1500	0.6020	0.5623	0.7495	0.6367	5	8
Trial 2 (second iteration)	1.5	7	5643	753	1506	0.7014	0.5291	0.7489	0.6554	8	8
Trial 2 (third iteration)	1.5	7	5589	771	1542	0.7181	0.1944	0.7519	0.5460	7	5
Trial 3	1.5	8	5631	757	1514	0.3939	0.7135	0.7087	0.6017	5	7
Trial 3 (second iteration)	1.5	8	5628	758	1516	0.4500	0.5556	0.7477	0.5765	5	2
Trial 3 (third iteration)	1.5	8	5667	745	1490	0.5094	0.6625	0.7221	0.6309	4	7
Trial 4	1.55	6.85	5643	753	1506	0.6470	0.4105	0.8036	0.6182	8	7
Trial 4 (second iteration)	1.55	6.85	5637	755	1510	0.6503	0.3047	0.7341	0.5603	9	6
Trial 4 (third iteration)	1.55	6.85	5643	753	1506	0.9309	0.0165	0.8070	0.5936	8	4
Trial 5	2	8	5607	765	1530	0.7462	0.3474	0.6911	0.5882	5	4
Trial 5 (second iteration)	2	8	5637	755	1510	0.6866	0.4361	0.6623	0.6000	10	5
Trial 5 (third iteration)	2	8	5655	749	1498	0.7725	0.3227	0.7255	0.5981	7	7
Trial 6	1	10	5649	751	1502	0	0.8983	0.6825	0.5746	8	2
Trial 6 (second iteration)	1	10	5610	764	1528	0.1375	0.8774	0.5672	0.5910	7	6
Trial 6 (third iteration)	1	10	5694	736	1472	0.1384	0.8920	0.6121	0.5965	6	3

In the Trial 1, after a complete scroll of 245 combinations of ANN architecture and training sets (based on previously described inner loops), we obtained a neural network with N_{h1}=8 and respectively N_{h2}=7 neurons on hidden layers. The

initial volume of data was randomly split in three parts: N_{tr}=5622, N_{val}=760 and N_{test}=1520 respectively. Each of these numbers is a sum of the values from each class (according with Table I). In Table II, the first line of values shows the main results of this described approach. Here, as previously mentioned, we denote by B_{12} the border point between the first and the second class and with B_{23} the next border point, which splits the second and the third class (the same as in Table I). Several trials, which are shown in Table II, use different values of these borders. The probability of the correct results in the each class, during the test phase, is separately represented in the same table, including the global probability of the correct results on the test set. As previously mentioned, each trial, with the same configuration of borders, is repeated two times more, from scratch, with other random initialization of training-validation-test sets, to see if the probabilities of correct classification are not a result of simply chance. This way we can check the repeatability of the results. It is obvious to see in Table II, that the volumes of data for each set (N_{tr}, N_{val}, and N_{test}), even if were randomly chosen the data, didn't vary too much in order to affect the percentages of theirs distributions.

During the second trial we established new borders between classes, having B_{12}=1.5 and B_{23}=8. After first and second iterations, the results were slightly better than those obtained after trial 1. Remember that the main idea was to equally distribute the volume of data between classes during the training process. In order to have a broad view we changed (but not quite dramatically) the borders on each trial. As a consequence, in next trial (Trial 3), there is a change of B_{23} (between second and third classes), but without visible improvement. Some similar behaviors we can appreciate in the next two trials (4 and 5). Condition has worsened, in trial 6, when we moved even farther the second border.

It became obvious that by moving to the left of the first border (B_{12}) we cannot expect to improve the results since it is obvious that the probability from the first class will decrease dramatically. As a consequence of these results it seems that we have to choose some values of the borders somehow similar with those from trials 1, 2, and 3 or even from the trials 4 and 5. Note that even if the trial 4 shows in Table I one of the best distributions of the elements, this is not automatically implying a better result.

Extending the numbers of borders in order to obtain four (or even more) classes wasn't a good choice for the model. In Table III, we selected one of the best trials (denoted here with number 7) in which we had split the data in four classes.

TABLE III. USING A MODEL WITH THREE BORDERS (B_{12}, B_{23} AND B_{34}) BETWEEN CLASSES

	B_{12}	B_{23}	B_{34}	$P_{Class\ 1}$	$P_{Class\ 2}$	$P_{Class\ 3}$	$P_{Class\ 4}$	P_{Test}	N_{h1}	N_{h2}
Trial 7	0.4	4	10	0	0.9396	0.0315	0.8172	0.4657	5	2
Trial 7 (second iteration)	0.4	4	10	0.5163	0.5875	0.2332	0.7216	0.5073	6	6
Trial 7 (third iteration)	0.4	4	10	0.2643	0.6286	0.3683	0.7508	0.4940	6	4

But, as we can see in Table III, the probabilities, to fall in one out of those four classes, have varied dramatically. Even if the second class and the fourth one have got better probabilities, the global probability of correct classification is around 0.5. It is quite a challenge to choose those numbers that separate successive classes. We had checked several combinations of borders but the results were even worse.

When more data will be available then we will include the adaptive retraining procedure [53] [54] in the model. This will allow us to refine the model when newest data are available. It will be an adaptation in which the old knowledge are not completely forgotten and the classification system will perform better as long as it progressively acquires more experience.

Furthermore, in order to validate the results obtained by the neural classifier, we were used statistical regression models. The strong influence of SMEs on macroeconomic results is well described by the correlation between the dynamic of the GDP, as macro result indicator, and the rate of new firm creation, as an indicator of the initiative spirit and the business environment. The creation of a new firm is a process monitored by the National Institute of Statistics, through a survey-based research harmonized by EUROSTAT. The rates of firm creation are computed as comparisons between the number of newly created firms and the existing stock in 1995 (388.180 enterprises).

The rate of new firm creation, always in the range of 13-17% until 2001, registered a significant increase afterwards, an increase generated by the improvement of the business environment, which led to an increase of the national production, as expressed by GDP. The number of newly created firms after 2003 is greater than 100.000, peaking in 2005 and oscillating afterwards. We remark a parallel evolution of GDP computed towards 1995 (IPIB) and the rate of new firm creation (RCNF), described by the equation estimated based on data for 17 years:

$$IPIB_i = 72.939 + 1.827\ RCNF_i \qquad (4)$$

The results of estimating the regression equation (4) are valid statistical, t-test probabilities of being lower than the level of significance ($P < \alpha = 0.05$). The parameter value estimated explanatory variable and indicated that a 1% increase in the rate of new firm creation leads to an average increase of 1.83% of GDP compared to 1995. Tests confirm the validity of the regression model ($F = 43.24$, $t = 6.54$, $P_{value} = 0.000009$). The variation in the GDP dynamic is explained to a large extent of 73%, by the variation in the creation of new private firms (Adjusted **R** Square = 0.73), while the covariance is 206.34. The equality between the correlation and the linear correlation coefficient confirms the correctness of the linear regression model (Multiple R and Pearson Correlation = 0.86).

The hypothesis of normal distribution of the residual variable is verified by the computations and the analyses of the statistics referring to its distribution, the value of Skewness being close to 0 while the value of Kurtosis is 2.2, thus indicating an almost symmetrical and slightly arched distribution, the Jarque-Bera test also indicating that this distribution is relatively normal [55], by the fact that the

probability associated to the test is higher than the chosen significance threshold ($0.8 > \alpha = 0.05$), which leads to the acceptance of the null hypothesis that the distribution is normal. Also, the hypothesis regarding errors' homoscedasticity was verified and confirmed [56].

The depth of the analysis on influencing factors is given by the attempt to identify the causes behind SME performance. Exogenous variables used in the regression models have been previously defined.

TABLE IV.

ANOVA[b] for RPC					
Model	Sum of Squares	df	Mean Square	F	Sig.
1 Regression	2.342	12	0.195	72.555	0.000[a]
Residual	21.219	7888	0.003		
Total	23.561	7900			
a. Predictors: (Constant), NMP, DPA, CT, DSA, IMP, CHF, CHP, CHE, AIT, ACT, CPT, TAMDC					
b. Dependent Variable: RPC ; Excluded Variables: CHT					

All data are taken from the financial reports of medium-sized firms (with 50-249 employees). The endogenous variable, the commercial profitability rate (RPC), was determined as percentage between the operating profit and total turnover. In preparing the data, all variables were normalized using two methods [57]. In the first case, we used the minimum and maximum, transposing the values in an interval, e.g. 0 - 1, or -1 to 0. In the second case, we used the Z score, normalizing the values around the average, with the standard deviation as the alternative. We established that between the exogenous and endogenous variables, there are no significant correlations, the Pearson correlation coefficient being between 0.19 and -0.19 [58]. The results of the regression model which involve all independent variables (Table IV), although statistically significant ($F=72.555$; $p<0.0005$), show the fact that the exogenous variables taken into consideration explain to a small extent the endogenous variable (R square=0.099).

Next, eliminate variables with little influence and composed a reduced model with the following independent variables (CHP-personnel expenditure, operating expenses - CHE, financial expenses - CHF and profit tax-IMP), the results are relatively similar.

TABLE V.

Coefficients[a]						
Model	Unstandardized Coefficients		Standardized Coefficients		t	Sig.
	B	Std. Error	Beta			
1 (Constant)	0.034	0.001			26.048	0.000
CHP	-0.047	0.011	-0.051		-4.325	0.000
CHE	-0.865	0.041	-0.325		-21.284	0.000
CHF	0.074	0.029	0.030		2.576	0.010
IMP	0.774	0.027	0.423		28.639	0.000
a) Dependent Variable: RPC ; Excluded Variables: CHT						

The model is statistically significant ($F=210.013$; $p<0.0005$), but the variables give a weak explanation of the commercial profitability rate (R square=0.096). The coefficients (Table V) are statistically significant different from zero (p<0.05), and the general form of the model which estimates the commercial profitability rate is:

$$RPC = 0.034-0.047CHP-0.865CHE+0.074CHF+0.774IMP \quad (5)$$

IV. CONCLUSIONS

The analysis made based on the data for Romanian SMEs shows that the economy has consolidated its character of market economy, with a high weight of the private sector in the national production. The model with the best validation shows the strong link between the rate of new firm creation and the GDP growth, explaining both the decrease of the national production when the crisis initiated and the worsening of the business environment.

The major influence, exerted by the SMEs sector evolution on national production, is proved by the regression equation between the rate of new companies' creation and the GDP growth, for the period 1995-2013, which indicate a strong dependence; also the regression coefficient confirms a strong impact of SMEs sector growth. Opposite to an economic common sense, the balance sheet indicators (assets, debts, financial expenses, personnel expenses, and income taxes) and even the number of employees, exert just a modest influence over the medium-sized companies' performance. The dependencies are explained by valid regression models, but with low values for the regression coefficients. This situation may be understood as a consequence of a weak economic mechanism, barely efficient, the medium-sized enterprises performance being determined mainly by off-balance sheet factors.

Governmental stimulation of new firm creation through: support for young entrepreneurs, grants for differentiated fiscal facilities in order to stimulate initiative in activities which generate added value, elimination of tax on reinvested profit, improvement of loans system for SMEs, creation of risk funds, improving the education of young entrepreneurs by including entrepreneurship disciplines in school programs, support in the international expansion of small firms, decrease of the tax level for small enterprises, tax exemption for firms that invest higher amounts than the value of the potential tax due, improving the legislations for public-private partnerships in order to attract SMEs to governmental investments, improvement of fiscal measures for firms created in the rural environment.

We have to remember that the selection of the borders between successive classes was intuitively suggested by specialists in economics and we saw the result of this approach. The total number of possibility is practically infinite and the searching for an optimal solution could be improved by using a genetic algorithm starting with a set of population defined by these trials, combined with constructing an ANN structure for each of them.

The neural classification system was able to identify the classes of profitability rate as long as theirs number is no more than three. As a conclusion of these results, it is difficult to extract a higher classification model by using these data. It seems that there might be necessary to find further information in order to obtain a better separation between classes. New

inputs with economic relevance could improve the results. Our future goal is to include in the neural model a retraining mechanism which implies a huge amount of historical data from previous years. This way it will finally result not only an intelligent system but also an adaptive one, which can be easily retrained on successive predefined intervals of time.

V. REFERENCES

[1] Commission of the European Communities, A "Small Business Act" for Europe (SBA), 2008; http://ec.europa.eu/enterprise/ policies/sme/small-business-act/files/sba_review_en.pdf

[2] Post-Privatization Foundation http://www.postprivatizare.ro/ romana/wp-content/uploads/2013/06/Raport-IMM-2013 (http://www.aippimm.ro/2013)

[3] I.M. Dragan, and A. Isaic-Maniu, "A Barometer Of Entrepreneurial Dynamics Above The Crisis", Romanian Statistical Review, vol. 61(7), 2013, pp. 53-64.

[4] D. Pîslaru, I. Modreanu, and Fl. Cîţu, Contribuţia IMM-urilor la creşterea economică: prezent şi perspective, Editura Economică, Bucureşti, 2012.

[5] M. Dragan, Al. Isaic-Maniu, "Some Characteristics of the New Entreprises and the Profile of New Entrepreneure," Revista de Management Comparat, vol.10, pp. 681-690, 2009.

[6] OECD-Organisation for Economic Co-operation and Development (2010). The Impact of the Global Crisis on SME and Entrepreneurship Financing and Policy Responses, OECD Paris.

[7] Irina-Maria Dragan, Al. Isaic-Maniu, Performance of small and medium enterprises, LAP Lambert Academic Publishing, Germany, 2012.

[8] Al. Isaic-Maniu and O. Nicolescu, "Characterization of performance differences between SMEs from processing industry in Romania", the 6th International Conference on Management of Technological Changes, Alexandroupolis, Greece, September 2009.

[9] O. Nicolescu, Al. Isaic-Maniu, and I. Dragan, White Charter of Romanian SMEs, Editura Sigma, Bucureşti, 2014.

[10] M. Mateev and Y. Anastasov, "Determinants Of Small And Medium Sized Fast Growing Enterprises In Central And Eastern Europe: A Panel Data Analysis," Financial Theory and Practice, 34 (3), pp. 269-295, 2010.

[11] M. Ayyagari, T. Beck, and A. Demirguc-Kunt, "Small and Medium Enterprises Across the Globe," Small Business Economics, 29 (4 December), pp. 415-434, 2007.

[12] Bulgarian Small and Medium Enterprises Promotion Agency, Analysis of the Situation and Factors for Development of SMEs in Bulgaria: 2011-2012, NOEMA, Sofia, 2012.

[13] J.E. Jennings, P.D. Jennings, and R. Greenwood, "Novelty and new firm performance: The case of employment systems in knowledge-intensive service organizations," Journal of Business Venturing, vol. 24, no. 4, pp. 338-359, 2009.

[14] H. Zhou and G. de Wit, "Determinants and dimensions of firm growth," SCALES-Scientific Analysis of Entrepreneurship and SMEs, Zoetermeer, Netherlands, 2009.

[15] I. Lejárraga, H. Lopez, H. Oberhofer, S. Stone, and B. Shepherd, "Small and Medium-Sized Enterprises in Global Markets: A Differential Approach for Services?", Doc. No. 165, OECD Trade Policy Papers from OECD Publishing, 2014.

[16] A. Heshmati, "On the growth of micro and small firms: evidence from Sweden," Small Business Economics, 17 (3), pp.213-228, 2001.

[17] H.T. Keh, T.T.M. Nguyen, and H.P. Ng, "The effects of entrepreneurial orientation and marketing information on the performance of SMEs," Journal of Business Venturing, vol. 22, no. 4, pp.453-612, 2007.

[18] Q.A. Subhan, R.M. Mehmood, and A. Sattar, "Innovation in Small and Medium Enterprises (SME's) and its impact on Economic Development in Pakistan," Proceedings of 6th International Business and Social Sciences Research Conference 3-4 January, 2013, Dubai, UAE, 2013.

[19] S. Evans, "The relationship between firm growth, size and age: estimates for 100 manufacturing industries," The Journal of Industrial Economics, 35 (4), pp.567-581, 1987.

[20] S.R. Henzel and M. Rengel, The macroeconomic impact of economic uncertainty: A common factor analysis, Munich: IFO Institute, 2013.

[21] Z. Tang, J.T. Tang, "Stakeholder–firm power difference, stakeholders' CSR orientation, and SMEs' environmental performance in China," Journal of Business Venturing, vol. 27, no. 4, pp. 436-455, 2012.

[22] M. Terziovski, "The relationship between networking practices and business excellence: a study of small to medium enterprises (SMEs), 2003 Measuring Business Excellence, Volume 7, Number 2, pp. 78-92, 2003.

[23] D. Hussey and P. Eagan, "Using structural equation modeling to test environmental performance in small and medium-sized manufacturers: can SEM help SMEs?", Journal of Cleaner Production, 15(4), pp. 303-312, 2007.

[24] F. Perri, A. Russo, and A. Tencati, "CSR Strategies of SMEs and Large Firms, 2007 Journal of Business Ethics, Volume 74, Issue 3, September, pp. 285-300, 2007.

[25] A. Islam, "Economic growth and crime against small and medium sized enterprises in developing economies," Small Business Economics, Springer, vol. 43, no. 3, pp. 677-695, 2014.

[26] S.K. Tayebi, S.M. Razavi, and Z. Zamani, "Effect of Financial Crisis on SMEs' Bilateral Trade Relations in Asia," Journal of Global Entrepreneurship Research, vol.2 (1), pp.1-16, 2012.

[27] J. Elliott, K. Ikemoto, "AFTA and the Asian Crisis: Help or Hindrance to ASEAN Intra-Regional Trade?", Asian Economic Journal, 18 (1), pp.1-23, 2004.

[28] K. Terdpaopong, (2011). Identifying an SME's debt crisis potential by using logistic regression analysis. RJAS- Rangsit Journal of Arts and Scieces Vol. 1 (1), pp. 17-26.

[29] J. Watson and S. Robinson, "Adjusting for risk in comparing the performances of male- and female-controlled SMEs", Journal of Business Venturing,vol. 18, no. 6, pp. 773-788, 2003.

[30] G.P. Zhang, "A computational study on the performance of artificial neural networks under changing structural design and data distribution," IEEE Transactions On Systems, Man, And Cybernetics - Part C: Applications And Reviews, vol. 30, no. 4, November 2000, pp. 451-462.

[31] P.C. Pendharkar, "A computational study on the performance of artificial neural networks under changing structural design and data distribution," European Journal of Operational Research, vol. 138, no. 1, April 2002, pp. 155-177, 2002.

[32] W.H. Beaver, "Financial ratios as predictors of failure, empirical research in accounting: selected studies," Journal of Accounting Research, vol. 4, pp. 71-111, 1966.

[33] E.I. Altman, "Financial Ratios, Discriminant Analysis and the Prediction of Corporate Bankruptcy," The Journal of Finance Vol. 23, No.4, pp. 589-609, 1968.

[34] R.A. Fisher, "The use of multiple measurements in taxonomic problems," Annals of Eugenics vol. 7, pp. 179-188, 1936.

[35] R.O. Edmister, "An Empirical Test of Financial Ratio Analysis for Small Business Failure Prediction," Journal of Financial and Quantitative Analysis, Vol. 7, No. 2, March 1972, pp. 1477-1493, 1972.

[36] H. Frydman, E.I. Altman, and D.L. Kao, "Introducing Recursive Partitioning for Financial Classification: The Case of Financial Distress," The Journal of Finance, vol. XL, no. 1, March 1985, pp. 269-291, 1985.

[37] L. Breiman, J.H. Friedman, R.A. Olshen, C.J. Stone, Classification and Regression Trees, Wadsworth Statistics/Probability, Chapman and Hall/CRC, Florida, 1984.

[38] J.R. Quinlan, "Induction of Decision Trees," Machine Learning, vol. 1, pp. 81-106, 1986.

[39] T. Kohonen, "Self-Organized Formation of Topologically Correct Feature Maps," Biological Cybernetics, vol. 43, no. 1, pp. 59–69, 1982.

[40] B.D. Ripley, "Neural Networks and Related Methods for Classifications," Journal of the Royal Statistical Society. Series B (Methodological), Vol. 56, No. 3, pp. 409-456, 1994.

[41] I. Nastac, A. Bacivarov, and A. Costea, "A Neuro-Classification Model for Socio-Technical Systems," Romanian Journal of Economic Forecasting, Vol. XI, No. 3/ 2009, pp. 100-109, 2009.

[42] A.L. Corcoran and S. Sen, "Using real-valued genetic algorithms to evolve rule sets for classification," Evolutionary Computation, Proceedings of the First IEEE World Congress on Computational Intelligence, June 1994, Orland, FL, Vol. 1, pp. 120-124, 1994.

[43] C. Cortes and V. Vapnik, "Support-vector networks," Machine Learning, September 1995, volume 20, issue 3, pp. 273-297, 1995.

[44] D. Michie, D.J. Spiegelhalter, and C.C. Taylor, Machine Learning: Neural and Statistical Classification, Overseas Press, 2009.

[45] C.H. Chou, C.C. Lin, Y.H. Liu, and F. Chang, "A prototype classification method and its use in a hybrid solution for multiclass pattern recognition," Pattern Recognition, Vol. 39, Issue 4, April 2006, pp. 624–634.

[46] A. Costea and I. Nastac, "Assessing the Predictive Performance of ANN Classifiers Based on Different Data Preprocessing Methods," Intelligent Systems in Accounting, Finance and Management, vol. 13, issue 4 (December 2005), pp. 217-250, 2005.

[47] O. Kurama, P. Luukka, and M. Collan, "Credit Analysis Using a Combination of Fuzzy Robust PCA and a Classification Algorithm", Chapter in Scientific Methods for the Treatment of Uncertainty in Social Sciences, J. Gil-Aluja et al. (eds.), Advances in Intelligent Systems and Computing 377, Springer International Publishing Switzerland 2015, pp. 19-29.

[48] I. Nastac, I.M. Dragan, and A. Isaic-Maniu, "Estimating Profitability Using a Neural Classification Tool", IEEE Proceedings of NEUREL 2014, Belgrade, Serbia, 25-27 November 2014, pp. 111-114.

[49] J.E. Jackson, A user guide to principal components, John Wiley, New York, 1991.

[50] I.T. Jolliffe, Principal component analysis, 2nd edition, Springer, New York, 2002.

[51] M.F. Moller, "A scaled conjugate gradient algorithm for fast supervised learning", Neural Networks, vol. 6, 1993, pp. 525-533.

[52] M.T. Hagan, H.B. Demuth, and M.H. Beale, Neural Networks Design, MA: PWS Publishing, Boston, 1996.

[53] D.I. Nastac, "An Adaptive Retraining Technique to Predict the Critical Process Variables," TUCS Technical Report, No. 616, June 2004, Turku, Finland, 2004. Available at: http://tucs.fi/publications/view/? pub_id= tNastac04a

[54] I. Nastac, "An adaptive forecasting intelligent model for nonstationary time series," Journal of Applied Operational Research, vol. 2, no. 2, 2010, pp. 117–129. Available at: http://www.orlabanalytics.ca/jaor/ archive/v2n2/jaorv2n2p117.pdf

[55] A.K. Bera and C.M. Jarque, "Efficient tests for normality, homoscedasticity and serial independence of regression residuals: Monte Carlo evidence," Economics Letters, 7(4): 313–318, 1981.

[56] S. Breusch and R. Pagan, "Simple test for heteroscedasticity and random coefficient variation," Econometrica, 47(5), pp. 1287–1294, 1979.

[57] G.J. Myatt, Making Sense of Data: A Practical Guide to Exploratory Data Analysis and Data Mining, John Wiley & Sons, 2007.

[58] D. Howitt and C. Duncan, Introduction to Statistics in Psychology, Prentice Hall, 2011.

Agent-based model of mosquito host-seeking behavior in presence of long-lasting insecticidal nets

Anna Shcherbacheva, Heikki Haario
School of Engineering Science
Lappeenranta University of Technology
Lappeenranta, Finland
Email: anna.shcherbacheva@lut.fi, heikki.haario@lut.fi

Gerry Killeen
Environmental Health and Ecological Sciences Thematic Group
Ifakara Health Institute
Ifakara, Morogoro, United Republic of Tanzania
Vector Biology Department
Liverpool School of Tropical Medicine
Pembroke Place
Liverpool, UK
Email: gkilleen@ihi.or.tz

I. BACKGROUND AND MOTIVATION

The efficiency of spatial repellents and long-lasting insecticide treated nets (LLINs) is a key research topic in controlling malaria. Insecticidal nets reduce the mosquito-human contact rate and simultaneously decrease mosquito populations. However, LLINs demonstrate dissimilar efficiency against different species of malaria mosquitoes.

Various factors have been proposed as an explanation, including differences in insecticide-induced mortality, flight characteristics, or persistence of attack. Here we present a discrete agent-based approach which enables examining the efficiency of LLINs, baited traps and Insecticide Residual Sprays (IRS). The model is calibrated with hut-level experimental data to compare the efficiency of protection against two mosquito species: *Anopheles gambiae* and *Anopheles arabiensis*.

We show that while such available data do not allow an unequivocal discrimination of all the optional factors, the overall protection efficiency can be estimated,and the model simulations can be generalized to community-scale scenarios to estimate the vectorial capacity under various degrees of protection.

II. SPECIFIC AIMS AND RESEARCH DETAILS

The aim of this paper is to present a numerical simulation framework that allows one to test various hypothesis concerning the impact of LLINs to malaria vector mosquitoes. Especially, we want to model and calibrate the difference in the behavior between *An. arabiensis* and *An. gambiae* as observed in hut experiments. A population of host-seeking mosquitoes is treated as a number of non-interacting agents, driven by external factors. We assume four basic driving effects: the attraction of odor emitted from a host (e.g., CO_2) and sensed by mosquitoes (see [2], [6], [5], [10], [14], [17]), the repulsion by physical net barrier and avoidance of a repellent, and the killing effect of the chemicals.

According to the data acquired from hut trials in [9], mortality of *An. arabiensis* is consistently lower than that of

An. gambiae or *An. funestus*. However, one can propose a number of plausible explanations of the underlying difference in host-seeking behavior of the species. A direct comparison, available from [16], suggests that *An. arabiensis* is a faster feeder than *An. gambiae*, which means that the former spends less time in contact with the skin (or net surface); hence, *An. arabiensis*'s exposure to treatment is shorter, with less dosage of chemical consumed. Another explanation proposes that *An. gambiae* and *An. arabiensis* exhibit different hunting patterns, with a tendency of the *An. gambiae* to stay close to the human-baited net (more 'determined' hunting), while *An. arabiensis* exhibits more random walk, also near the net. In the model simulation, this would be achieved by more direct movement towards the human. One more alternative explanation suggests that *An. gambiae* and *An. arabiensis* feature different persistence of blood-feeding attempts, since odorant receptors of anthropophilic *An. gambiae* are narrowly tuned to compounds of human sweat, see [11], [15].

All the above described factors can be implemented in the modeling approach presented here. Each factor needs to be given by some parametric formula, with the parameters calibrated against measurement data. This calls for parsimonious models where the expressions for the basic factors are given with formulas containing a minimal number of parameters.

We show how to fit such models to the data of [9]. The identifiability of the parameters is studied using extensive Monte Carlo (MCMC) sampling methods. It turns out the data in [9] does not allow a unanimous identification of all the possible factors, as different model variants are able to give equally good fits. We demonstrate that they, on the other hand, lead to essentially the same models for the overall protection efficiency. This enables us to model the reduced mosquito–human contact rates and increased mosquito mortality in community-scale scenarios consisting of several households under various degrees of protection.

Agent based approaches have been earlier applied for other purposes. The results of such models highlight, for instance, the role of heterogeneity in host movement, mosquito distribution and density, and the environment in mosquito-borne disease spread, [12],[1], [3], [7], [8], [13]. In [4] a combination of continuous modeling and agent based simulations was presented in order to simulate the flight of mosquitoes towards the host in outdoor conditions including wind. The aim of this work is different, to characterize the impact of LLINs to host-seeking behavior of different mosquito species. We restrict here the model calibration to the situation of mosquitoes and a host in a hut, but the approach can be easily modified for different experimental conditions. Also, the calibrated model can be used to simulate the vector dynamics in more complex situations, in domains lager in space and time, in combination with continuous modeling such as in [4], and with larger host populations with varying levels of protection.

REFERENCES

[1] S.J. Almeida, R.P. Ferreira, Á.E. Eiras, R.P. Obermayr, and M. Geier. Multi-agent modeling and simulation of an *Aedes aegypti* mosquito population. *Environ Model Softw*, 25(12):1490–1507, 2010.

[2] M.F. Bowen. The sensory physiology of host-seeking behavior in mosquitoes. *Annu Rev Entomol*, 36:139–158, 1991.

[3] D.L. Chao, S.B. Halstead, M.E. Halloran, and I.M. Longini. Controlling dengue with vaccines in thailand. *PLoS Negl Trop Dis*, 6(10):e1876, 2012.

[4] B. Cummins, R. Cortez, I.M. Foppa, J. Walbeck, and J.M. Hyman. A spatial model of mosquito host-seeking behavior. *PLoS Comput Biol*, 8(5):e1002500, 2012.

[5] T. Dekker and R.T. Cardé. Moment-to-moment flight manoeuvres of the female yellow fever mosquito (*Aedes aegypti L.*) in response to plumes of carbon dioxide and human skin odour. *J Exp Biol*, 214(20):3480–3494, 2011.

[6] T. Dekker, M. Geier, and R.T. Cardé. Carbon dioxide instantly sensitizes female yellow fever mosquitoes to human skin odours. *J Exp Biol*, 208:2963–2972, 2005.

[7] P.A. Eckhoff. A malaria transmission-directed model of mosquito life cycle and ecology. *Malar J*, 10:1–17, 2011.

[8] W. Gu and R.J. Novak. Agent-based modelling of mosquito foraging behaviour for malaria control. *Trans R Soc Trop Med Hyg*, 103(11):1105–12, 2009.

[9] J. Kitau, R.M. Oxborough, P.K. Tungu, J. Matowo, R.C. Malima, S.M. Magesa, J. Bruce, F.W. Mosha, and M.W. Rowland. Species shifts in the *Anopheles gambiae* complex: Do llins successfully control *Anopheles arabiensis*? *PloS ONE*, 7(3):e31481, 2012.

[10] M. Lehane. *The Biology of Blood-Sucking in Insects, Second Edition*. NY: Cambridge University Press, 2nd ed edition, 2005.

[11] L.M. Lorenz, A. Keane, J.D. Moore, C.J. Munk, L. Seeholzer, A. Mseka, E. Simfukwe, J. Ligamba, E.L. Turner, L.R. Biswaro, F.O. Okumu, G.F. Killeen, W.R. Mukabana, and S.J. Moore. Taxis assays measure directional movement of mosquitoes to olfactory cues. *Parasit Vectors*, 6:131, 2013.

[12] C.A. Manore, K.S. Hickmann, J.M. Hyman, I.M. Foppa, J.K. Davis, D.M. Wesson, and C.N. Mores. A network-patch methodology for adapting agent-based models for directly transmitted disease to mosquito-borne disease. *J Biol Dyn*, 9(1):52–72., 2015.

[13] H. Padmanabha, Fabio Durham, David orrea, M. Diuk-Wasser, and A. Galvani. The interactive roles of *Aedes aegypti* super-production and human density in dengue transmission. *PLoS Negl Trop Dis*, 6(8):e1799, 2012.

[14] Z. Syed and W.S. Leal. Acute olfactory response of *Culex* mosquitoes to a human-and bird-derived attractant. *Proc Natl Acad Sci U S A*, 106(44):18803–8, 2009.

[15] I.V. van den Broek and C.J. den Otter. Olfactory sensitivities of mosquitoes with different host preferences (*Anopheles gambiae s.s., An. arabiensis, An. quadriannulatus, An. m. atroparvus*) to synthetic host odours. *J Insect Physiol*, 45(11):1001–1010, 1999.

[16] J.A. Vaughan, B.H. Noden, and J.C. Beier. Concentrations of human erythrocytes by *Anopheline* mosquitoes (*Diptera: Culicidae*) during feeding. *J Med Entomol*, 28(6):780–786., 1991.

[17] N.J. Vickers. Mechanisms of animal navigation in odor plumes. *Biol Bull*, 198(2):203–212, 2000.

A multi-criteria decision-making tool with information redundancy treatment for design evaluation

Nikolai Efimov-Soini*, Mariia Kozlova, Pasi Luuka, Mikael Collan
School of business and management
Lappeenranta University of Technology
Lappeenranta, Finland
*spb2010@mail.ru

Abstract — **This article introduces a multi-criteria decision making (MCDM) method that takes into account interdependency of criteria. Recognizing information redundancy in related criteria the proposed approach offers their weight formation based on their interaction. Further, normalized estimated criteria and their weights are aggregated by means of Fuzzy Heavy Weighted Averaging (FHWA) operator. The approach is illustrated with a numerical case study of engineering design selection problem.**

Keywords— Multi-criteria decision making, design assessment, fuzzy logic, information redunduncy

I. INTRODUCTION

The design evaluation is a crucial task on the conceptual design stage, as far as the concept chosen in this stage influences the whole further product life-cycle [1]. On the one hand, the information about concepts is often incomplete, uncertain and evolving. On the other hand, key decision criteria are often interdependent that hinders unbiased decision-making.

This paper proposes a method suitable for the design evaluation. In contrast to existing techniques [2], the distinct feature of the new method is that it takes into account information redundancy of evaluating criteria. The latter allows more accurate decision-making. In this paper, the flow meter case is used to illustrate the proposed method.

II. THE PROPOSED APPROACH

The proposed method is based on the procedure presented in [3] and incorporates in addition information redundancy treatment. It contains six main steps:

1. Transformation of data vector into fuzzy numbers to capture imprecision and uncertainty of estimates;

2. Normalization of fuzzy numbers in order to transform them into comparable units and enable aggregation;

3. Distinguishing between cost and benefit criteria and taking a complement from cost ones;

4. Creating weights by using interaction matrix of criteria;

5. Aggregation of the fuzzy vector with given weights by means of FHWA operator;

6. Forming rankings from resulting aggregated fuzzy numbers [4].

The interaction matrix in the step 4 is a special matrix that reflects dependence (or independence) of criteria to each other. The interaction matrix can be defined as:

$$Y = (y_{ij})_{n \times n} \qquad (1)$$

Where $y_{ij} \in \{0,1\}$ denoting presence/absence of interaction. We assume $y_{ii} = 0$, meaning that variable does not interact with itself. The interaction vector is created by cardinality of interactions presented:

$$I_j = \sum_{i=1}^{n} y_{ij} \qquad (2)$$

Further, it is scaled to a unit interval by

$$\hat{I}_j = \frac{I_j}{I_{max}} \qquad (3)$$

where $I_{max} = n$ denoting maximum possible interactions with variable. The weights are formed by taking the complement of the the interaction vector:

$$w_j = 1 - \hat{I}_j \qquad (4)$$

Thus, the weight vector reflects information redundancy in the criteria. The more the interdependence between one criterion with others, the lower its weight in the final estimate.

A Fuzzy Heavy Weighted Averaging (FHWA) operator [3] is used for aggregation. FHWA of dimension n is a mapping operator that maps Un → U that has an associated weighting vector W of dimension n such that the sum of the weights is between [1,n] and $w_i \in [0,1]$, then:

$$FHWA(\hat{a}_1, \hat{a}_2, \cdots, \hat{a}_n) = \sum_{i=1}^{n} w_i \hat{a}_i \qquad (5)$$

where ($\hat{a}_1, \hat{a}_2, \cdots, \hat{a}_n$), are now fuzzy triangular numbers

For details on other steps, see [5].

III. CASE ILLUSTRATION

The application of the above algorithm is illustrated with a selection problem among designs of three different flow meters: electromagnetic (EM), turbine and ultrasonic. The flow meters are used for measuring of the flow of liquids or gases in the different areas, such as construction, oil and gas, nuclear power etc. For these devices eight important criteria are identified: the cost of device, the work time, the electricity consumption, the accuracy, the amount of liquids on which they can operate, the easiness of installation, the processing of the electronic signal, and the shelf-time. The specifications on each flow meter type is presented in Table I [6].

TABLE I.   SPECIFICATIONS OF THE FLOW METER DESIGNS

	EM (Piterflow RS50)	Turbine (Okhta T50)	Ultrasonic (Vzloyt MR)
1. Cost, rubles	16150	4240	34800
2. Work time, hours	80000	100000	75000
3. Consumption, V*A	6	0	12
4. Accuracy, m3/hour	36±2%	30,00±2%	35±2%
5. Operatied liquids	Water, and sold water, dirty water	Only clear water	All liquids
6. Easiness of installation	Easy	Elementary	Average
7. Processing of the electronic signal	Yes	No	Yes
8. Shelf-time, years	4	5	2

Detailed computations in accordance with the defined algorithm are left outside this extended abstract, however, we dwell upon the weight formation here. Table II presents the interaction matrix of eight defined criteria numbered consequently.

TABLE II.   INTERACTION MATRIX

	1	2	3	4	5	6	7	8
1	0	1	0	1	1	0	1	0
2	1	0	0	0	0	0	0	0
3	0	0	0	0	1	0	1	0
4	1	0	0	0	0	0	0	0
5	1	0	1	0	0	1	1	0
6	0	0	0	0	1	0	1	0
7	1	0	1	0	1	1	0	0
8	0	0	0	0	0	0	0	0
Sum	4	1	2	1	4	2	4	0

The interaction matrix shows that e.g. cost of the design is influenced by its durability, accuracy, amount of operated liquids and ability to process the electronic signal. Thus, the overall interaction of the first criterion with others is equal to 4 (defined as cardinality). Scaling it to unit interval with the total possible interactions equal to 8 returns 0.5. Hence, the weight of this criterion, computed as the complement is equal to 0.5. The overall weight vector defined in this manner is [0.5 0.875 0.75 0.875 0.5 0.75 0.5 1].

Aggregation of normalized fuzzified criteria with their weights provides the final fuzzy estimates for each design represented on Fig. 1.

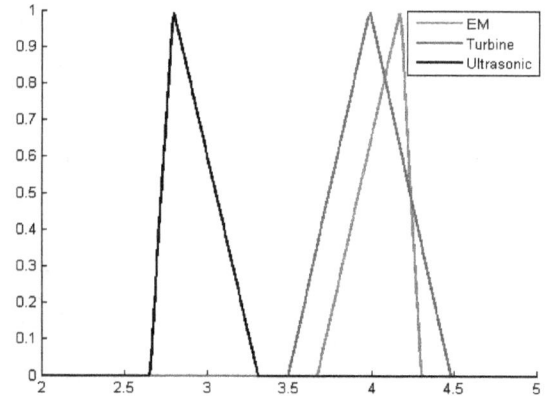

Fig. 1. Final fuzzy numbers for each flowmeter.

According to the Kaufmann and Gupta method [4] a unique linear order of fuzzy numbers can be found by using some properties of the numbers as criteria for ordering. In this method, the following three properties are used: removal number, mode and divergence. If the first criterion does not give a unique linear order, then the second criterion is used, and ultimately the third criterion is applied as well. In our case the first criterion is enough to obtain the final ranking. In particular, the removal for a triangular fuzzy number is calculated as:

$$\tilde{A} = \frac{a_1 + 2a_2 + a_3}{4} \qquad (6)$$

The final ranking (Table III) of these fuzzy estimates proposes electromagnetic flow meter as the best one, followed by turbine and ultrasonic designs.

TABLE III.   RANKINGS FOR EM, TURBINE AND ULTRASONIC DESIGNS

Flow meter	Final ranking
EM	1
Turbine	2
Ultrasonic	3

IV. CONCLUSIONS

This paper introduces a new multi-criteria decision-making method that treats information redundancy in evaluating criteria. The approach is illustrated on engineering design selection problem. The results show its effectiveness in removing the effect of interdependent criteria from the final estimate. The proposed approach is suitable for the majority of MCDM problems and can potentially find its application in a variety of industries and academic fields.

REFERENCES

[1] Ullman, David G. 2010 – The mechanical design process/David G. Ullman, 4th edition ISBN 978-0-07-297574-1-ISBN 0-007-297574 (alk. paper). McGraw-Hill series in mechanical engineering.

[2] Concept selection methods – a literature review from 1980 to 2008. G. Okudan, S. Tauhid. International Journal of Design Engineering, Vol. 1, No. 3, 2008, pages 243-277.

[3] Collan, M. and Luukka, P. (2015). Strategic R&D project Analysis: Keeping it simple and smart,Using fuzzy scorecards to collect data for strategic R&D projects & analyzing and seleting projects with a system that uses new fuzzy weighted averaging operators, Submitted

[4] Kaufmann, M. and M. Gupta (1985). Introduction to fuzzy arithmetics: theory and applications. New York, NY, Van Nostrand Reinhold.

[5] Efimov-Soini, N., Kozlova, M., Luuka, P. and Collan M. (unpublished). A multi-criteria decision-making tool with information redundancy

[6] Termotronic, product description. Available at: http://termotronic.ru/products/ [Accessed 12.05.2016]

Customer Need Based Concept Development

Kalle Elfvengren
Lappeenranta University of Technology
School of Business and Management
Lappeenranta, Finland
kalle.elfvengren@lut.fi

Mika Lohtander
Lappeenranta University of Technology
School of Energy Systems
Lappeenranta, Finland
mika.lohtander@lut.fi

Leonid Chechurin
Lappeenranta University of Technology
School of Business and Management
Lappeenranta, Finland
leonid.chechurin@lut.fi

Abstract—The front end of innovation presents one of the greatest opportunities for improving the overall innovation process. The front end is defined as the period between when an opportunity for a new product is first considered and when the product idea is judged ready to enter formal development. A big problem in the front end is that it is often ineffective, and a huge part of new ideas are deleted during the front end. There is a need for a systematic and practical procedure to make good quality concept propositions. This paper presents a framework to enhance the new product concept ideas. The paper reviews the possibilities of New Concept Canvas, QFD and TRIZ to enhance the product concepts in front-end planning.

Keywords—front end of innovation; concept planning; QFD; TRIZ

I. INTRODUCTION

There is a clear link between successful new innovation and the company's ability to fulfil critical customer needs. Successful manufacturers, such as Apple and Samsung show strong ability to understand market needs and create new products to meet these needs. However, idea management can be very challenging. For example, Nokia had a new and working prototype for a touchscreen smart phone long before the iPhone was introduced, but Nokia killed this new product concept because executives thought that it was too risky and they did not see the hidden market and customer need behind it [1].

It is not an easy task to pick up the winners from the new product concepts that are in the early planning phases. The resources are always scarce and hard decisions have to be made when the executives prioritize future product development efforts. There are a lot of different methods available for companies to support the front-end phases of the innovation process. There are methods and procedures for collecting and analyzing customer need data, and ideation methods to generate new product ideas. These methods include e.g. several brainstorming techniques, group work methods and toolsets such as TRIZ.

A one big problem in the front end is that it is often ineffective. Only very few preliminary concept plans lead to the market entry of a new breakthrough product. A great number of potential new product concepts are discarded during the early phases. Sometimes the concepts fail simply because they do not fit the company strategy or they are too radical.

The paper presents New Concept Canvas –tool (NCC) which help to enhance the potential new concepts and product ideas. The study also reviews the possibilities of combined use of NCC, Quality Function Deployment (QFD) and TRIZ as an effective and systematic process to invent new products that fulfil customer needs.

The research methodology used in this study is the constructive approach [2]. The logic in the constructive approach is to design a new construct and test its applicability in cases. The goal is to build new constructs that are tied to the current knowledge, and the results of the research are evaluated based on their newness and applicability. The authors of this study utilized the new product development success factor studies, and used their own practical experiences in the industry to develop NCC framework. The case-study presents an application made for a Finnish company which offers cooling device solutions for IT and electronic industry.

II. FRONT END OF INNOVATION

In order to develop processes and tools to support the front end activities, an important part of this research was to review the characteristics, challenges and success factors of the front end activities. The knowledge about managing the front end process was obtained by examining previous literature.

The front end of innovation is defined as those activities that take place prior to a formal new product development process. The front end ranges from the generation of an idea to either its approval for development or its termination [3]. The output of the front end is a detailed business plan. The front end activities strongly determine which projects will be executed, and the quality, costs and time frame are defined here [4].

A great amount of research exists on best practices for starting a new product development (NPD) process, e.g.[5][6]. A summary of NPD success factors can be found e.g. in a study by Conn [8]. Typical risks and uncertainties endangering the success of an innovation include for example failing to develop the technology as planned, inaccurate estimates of the future market demand, or a combination of both.

Many studies have noted the importance of front end activities (e.g. [9][10]). However, according to Khurana and Rosenthal [11], companies have identified the front end as being the greatest weakness in NPD. The front end of innovation offers great opportunities for improving the overall innovation process [12]. Successful execution of the front end is important for the success of the product on the market. An empirical study by Cooper and Kleinschmidt [13] showed that the biggest differences between winners and losers were found in the quality of the execution of front end activities. Two factors were identified as playing a major role in product success: the quality of executing the front end activities, and a well-defined product and project prior to the development phase [13]. Dwyer and Mellor [10] discuss the benefits of early stage business case and market need analysis. They balance this commercial focus with the need for an early-stage complimentary technical focus.

Robert Cooper's State-Gate process proposes that the front end consists of a single idea generation phase [7]. Cohen et al. [14] have expanded Cooper's idea generation stage into three new stages for managing technical uncertainty. However, managing technology is only part of front-end activities. Langerak et al. [15] have studied how critical predevelopment activities are for a market-oriented firm to achieve superior performance. The results provide evidence that market orientation is positively related to proficiency in idea generation/screening and strategic planning. Koen et al. [12] introduce a new concept development model for organizing the core activities in the front end of innovation, which increases the understanding and positive image of the front end activities.

III. THE IMPORTANCE OF UNDERSTANDING CUSTOMER NEEDS IN THE FRONT END

A large number of studies concerning the importance of identifying customer needs in the front-end have been carried out (e.g. [16][17][18][19]). These studies have searched for factors that make product development in some companies more successful than in others. Factors that have been pointed out in these studies are the significance of the early phases of product innovation and the assessment of market and customer needs for the success of NPD. Developing a successful product requires accurate understanding of customer needs. According to several studies [20][21][22], product development projects that are based on carefully defined customer needs are more likely to succeed than those based on new technological opportunities. The economic success of manufacturing companies depends on their ability to identify the needs of their customers and to quickly create products that meet these needs and can be produced at a low cost. The precious resources should be aimed at those development activities which increase customer satisfaction most. In addition, by carefully defining customer needs in the early phases of innovation process, large and time-consuming changes can be avoided at the later stages of the development work, which can significantly reduce the total time required by development activities [22]. To be useful, the information on customer needs has to be in a form that can be easily communicated, and it should also be considered useful and trustworthy by the persons in R&D who need it. Therefore, systematic and competent methods, such QFD can be very useful.

IV. THE NEW CONCEPT CANVAS FRAMEWORK

The authors have created a framework for further development of new product ideas. NPD success factor studies were utilized when the framework was planned. Identified critical success factors of product development activities are for example delivering a differentiated product with unique customer benefit, superior value, early understanding of customer needs and wishes, identified weaknesses of competitors' products, sharp and early product definition, and a target market definition. The framework consists of nine essential segments. Inspiration for the developed framework came from the well-known Business Model Canvas technique by Osterwalder and Pigneur [23]. The principle of the framework is that the new product idea is systematically developed further by filling the NCC framework. Table 1 presents the segments, the information needed for each segment, as well as justification of the importance of each segment by success factors.

TABLE 1. THE NCC FRAMEWORK CONSTRUCTION

Framework segment	Required information	Success factors
Problem	- Clarify and describe the problem - Point of view: customer's problem - What is the business opportunity?	- Customer focus - Solving the real problem
Solution	- Clarify and describe the solution - How will it work?	- Innovative solution
Key customer needs	- Clarify and describe the customer needs - Evaluate the importance of each need (1-5 scale)	- Understanding the right needs - Need pull
Customer value	- Why will the target customers buy our solution? - What needs will our solution satisfy and how? - What benefits will the customer get?	- Unique customer benefit - Superior value
Customer segments	- Who are our most important customers? - For whom are we creating value?	- Market orientation - Target market definition
Technical features	- Describe the main features of the solution - Describe also the technical features	- Sharp and early product definition
Competition	- Who are our main competitors? - What other solutions fulfill the same needs?	- Understanding the markets - Benchmarking competitors
How our solution will overcome the competitors?	- Why will our solution be better? - What needs are we going to fulfill better? - What features will be better?	- Superior products which stand out from competitors - Identifying the weaknesses of competitor products - Market/competition analyses
Needed technological know-how and risks	- What enabling technologies will we need? - Needed partners and resources - What are the risks for the success of the concept? - Manufacturing problems? Maintenance aspects?	- Technology testing - Risk awareness

Figure 2 illustrates the principles and visual look of the developed NCC framework. It has nine segments that need to be filled when the NPD-team is planning a preliminary concept idea. The titles of the segments have been written in different colors. This simply helps to identify the areas which are normally tied closely together (problem and solution, customer focus, technical aspects and competition related areas). For example, it helps to understand the real problem if the company has profound understanding of the customer needs. Different customer segments can also have different needs. Fulfilling the needs add directly to customer value. The desired technical features of the new concept also have a direct effect on the risks and on what kind of technological know-how the company needs for the new product.

Problem ?	Solution	Competition
• Clarify and describe the problem • Point of view: customer's problem • What is the business opportunity?	• Clarify and describe the solution • How will it work?	• Who are our main competitors? • What other solutions fulfill the same needs?
Key customer needs	Customer value	How will our solution overcome the competitors?
• Clarify and describe the customer needs • Evaluate the importance of each need (1-5 scale)	• Why will the target customers buy our solution? • What needs will our solution satisfy and how? • What benefits will the customer get?	• Why will our solution be better? • What needs are we going to fulfill better? • What features will be better?
Customer segments	Technical features	Needed technological know-how and risks?
• Who are our most important customers? • For whom are we creating value?	• Describe the main features of the solution • Describe also the technical features	• What technologies will we need? • Needed partners? • Needed resources? • Manufacturing problems? • Maintenance aspects

Fig. 1. Illustration of the NCC Framework.

The canvas has been used in the innovation management related course at Lappeenranta University of Technology, and the students who took part in the course were at their Master's level. During the course the students came up with several new product ideas in an brainstorming workshop. The next task was to choose the most promising or interesting new ideas, and to develop them further with NCC in three-person groups. The students felt that the NCC helped them to create a clear and visual presentation of the concept. The common feedback about the NCC was positive. According to students' comments, the canvas gave them "a tool to support the discussion" and "a simple documentation framework". They also stated that the canvas gave a good possibility for evaluate the concepts more democratically, because the format of all concepts were presented in a similar way.

On the basis of the experiences of the student cases, it can be concluded that the NCC has potential to be a needful tool in the front end of innovation.

V. QFD AND TRIZ IN CONCEPT PLANNING

The one target of this paper is to study using NCC, the House of Quality (the first phase of QFD) and TRIZ methods together as a process. These methods was chosen for this study because QFD supports customer need -based concept planning systematically, and TRIZ is used to create new ideas in a systematic and efficient way. Also after using the NCC, it seems to be natural to continue with the QFD matrix. NCC and QFD suits nicely together: output of NCC can be used as input to QFD.

QFD is a tool to translate the user requirements and requests into product designs. Shigeru Mizuno and Yoji Akao created QFD in Japan in the 1960s [24]. QFD links the customer needs with technical requirements from design, development, engineering, manufacturing, and service functions. The goal of QFD is to build a product that fulfills the customer needs, instead of offering a product that is technologically inspiring to the developer's engineer-eye [21].

House of Quality is typically the first phase of the QFD-process. In House of Quality, the QFD-matrix includes at least customer needs, technical product features and the correlations between needs and features.

TRIZ (Theoria Resheneyva Isobretatelskehuh Zadach) is a problem analysis and systematic ideation theory derived from the patterns of invention in patent data. The father of TRIZ was inventor Genrich Altshuller from the Soviet Union. He introduced the approach in 1956 [25], but the theory development started right after the Second World War [26]. TRIZ was developed as the analysis/solving methodology, when the problem is given in engineering terms (for example how to damp steering wheel vibrations or how to reduce door opening force). In other words, it was taken for granted that solving this very problem would improve the quality of the product, reduce the costs etc. It was a typical inventor's problem setup in the command-driven economy to listen to the Voice of the Product only. There are several systematic approaches bridging the customer's expectations with the product's engineering parameters. QFD seems to be the most popular instrument so far. Naturally, ideas to merge QFD and TRIZ appeared immediately as soon as TRIZ came onto the stage.

The role of TRIZ in new product development is important because it can help engineers to generate new ideas to overcome problems and to develop new product ideas. By using of QFD and TRIZ together, a company can be able to identify the most important customer need -related product features and then solve the emerging technical difficulties in the realization of these features with creative and ideal solutions.

There are several studies about utilizing QFD and TRIZ, and according to the literature, QFD and TRIZ can support each other. The article of Jugulum and Sefik seems to be among the first indexed papers on manufacturing management calling up TRIZ in the context of quality improvement methods like QFD [27]. In this article, TRIZ is mentioned and its potential is highlighted.

The study of Yamashina, Ito and Kawada [28] is one of the most often cited articles that presents a systematic way to integrate TRIZ and QFD. In their approach, QFD first turns the customer's requirements into quality engineering parameters, and then it reveals the negative correlation relationships between quality parameter pairs. This means that there is a contradiction between these two parameters, and the suggestion is to use TRIZ to resolve the contradiction.

Sakao [29] presents a methodology where the TRIZ stage of design is deeply integrated into QFD. It suggests using these tools together with Life Cycle Assessment (LCA) to harmonize the voices of the customer, environment and product. The method helps to reveal contradiction between quality characteristics and environmental considerations, and applies TRIZ to resolve them. QFD also serves as a contradiction determination tool in [30], where an LCA-based eco-design is in focus as well. A roadmap combining QFD, TRIZ and Kano methods is presented in [31]. As in most of the cases, QFD is used to define contradictions and TRIZ is used to resolve them and to generate new concepts. These concepts are evaluated by using Kano analysis.

The whole cycle of design, from product to manufacturing, is considered in [32]. The article suggest combining the QFD-TRIZ in all of its stages. A carefully described case study of a PC notebook is given, and environmental aspects are also taken into account by the approach. A fuzzy QFD is reported to be useful prior to TRIZ application in [33]. The fuzzy QFD does not provide any new insight, but it helps experts to give their opinion in a more comfortable (less certain) form. This approach is used for non-technical service design area.

A. Case Study: Thermal Handling Device

The case study presents a QFD-TRIZ application made for a Finnish spin-off company. The case company offers thermal handling device solutions for information technology and electronic industry. The company is developing a new type of thermal handling device, which provides small size and relatively good energy efficiency compared to the products in the market today. The most important feature is a new power class. The cost-efficient and compact device is suitable for all industrial applications whose cooling power rate is 50-500W and heating rate 10-100W. In many cases the new device suites for use in common office environments as well.

Optimal target applications for the system include electronics that need efficient cooling, are housed in tightly closed cases, or operate in dirty environments. The efficiency of the new system is based on the well-known Peltier technology, but the innovation is based on a new way of controlling the device. The new way to handle Peltier elements reduces for example the share of the total power that small data centers consume for cooling purposes. The packaging density of power electronics can also be increased with this technology. At the same time, when the power can be directed to actual processing and the power needed for cooling/heating is reduced, savings are made in production costs and total energy consumption.

The NCC framework was used firstly to identify the most important customer needs (through the problem/solution/value points of view), and also to identify the most important technical features and problems/risks. The QFD-matrix for the next version of the thermal handling device was done with the cooler expert in the case company. The House of Quality matrix was made in a two-hour meeting (see Appendix). The purpose was to identify what technical features of the new product are the most satisfying from the customer's point of view. Also, the goal was to identify the most important technical features and recognize the most difficult development tasks.

B. Applying TRIZ to the Case

By studying the QFD-matrix, it can be seen that the two most important technical features of the product are COP > 1 Coefficient of performance and Powerclass 100-200 W. These features are the ones of high importance: there are some strong correlations with important customer needs. QFD helps the managers to recognize quickly which features are the right ones to aim the development efforts at. It can also be seen in the matrix that these features have problems that make them very difficult to develop. At this stage, TRIZ could offer a systematical way to approach these problems.

Another way is to start from the engineering parameters; the QFD case matrix reveals negative correlation relationships between some quality parameter pairs. This means that there could be contradictions between these parameters, and TRIZ can be used to resolve these contradictions. The matrix has parameters like 'maximum weight is under 2 kg', 'impact resistance' and 'vibration resistance'. At the same time, the device should be very robust (weight usually brings along sturdiness and durability), so that it can resist vibrations and impacts. The contradictions are: minimum weight and maximum impact resistance, or minimum weight and maximum vibration resistance. The TRIZ can now be applied to generate ideas to solve these contradictions. E.g. the TRIZ tool '40 inventive principles', which reveals the principles that can be found in hundreds of thousands of patents [34], can be used in the contradiction analysis or on their own as an ideation tool. The contradictions to solve are Weight of stationary (feature to improve) and Durability of non-moving object (feature to preserve). The suggested inventive principles from the TRIZ matrix are Taking out, Cheap short living objects, Periodic action and Universality. These principles could offer a new ground for ideation to resolve the contradictions in this case.

VI. CONCLUSIONS

This paper concentrated on the important phase of new product creation: the front end of innovation. The significance of identifying the most important customer needs and problems was discussed through the literature. This paper has presented practical process to help to enhance the new product concepts. The presented process utilizes also the principles of QFD and TRIZ. The proposed new NCC framework has been tested with university students. The NCC framework seems to be a promising tool and it seems to offer a systematic approach for conceptual planning. It can also offer a company an official format for presenting raw ideas at front-end screenings. The template can be used e.g. in electronic idea bank of a company's NPD, and it fits well to the presented process together with QFD.

In the case-study part of the paper, the authors made a QFD-matrix for the development of a new two-phase cooling device. The QFD helped the company to organize the customer needs of their product. The customer needs were commonly known, but with the QFD the needs were organized and their importance was evaluated. QFD helped to concentrate on the needs and to make a summary of them. Also the technical parameters and targets were evaluated systematically and profoundly for the first time. QFD offered a clear way to evaluate the parameters from the customer need point of view and to set the right development targets. TRIZ offered an inventive approach to continue the concept development and to find new ideas to overcome the technical problems. QFD helped to find TRIZ contradictions and to show the most important (from the customer's point of view) and hardest technical development challenges. Because the TRIZ process is time-consuming, it is important to be able to focus the TRIZ activities on the most important problems.

Further research of best practices in the utilization of QFD and TRIZ together is planned to be done with other company cases. The scientific contribution of this research is still rather limited, but with additional real-world cases the validity of the findings can be increased.

QFD helps to analyze the interconnection of engineering parameters and define which are most responsible for quality improvement. Most studies recognize that the interconnection analysis is similar to contradiction detecting in TRIZ and therefore suggest a natural way of integration: QFD is followed by contradiction elimination technique to find a new paradigm for the product.

ACKNOWLEDGMENT

The authors are grateful for the Finnish Funding Agency TEKES that funded this research.

REFERENCES

[1] A. Troianovski and S. Grundberg, "Nokia's Bad Call on Smartphones", The Wall Street Journal, July 18, 2012.

[2] S. Mäkinen, "A Strategic Framework for Business Impact Analysis and its Usage in New Product Development", Acta Polytechnica Scandinavica, Industrial Management and Business Administration Series No. 2, Espoo, 1999.

[3] S. Murphy and V. Kumar, "The Front End of New Product Development - A Canadian Survey", R&D Management, Vol.27, No.1, pp.5-16, 1997.

[4] C. Herstatt, B. Verworn and A. Nagahira, "Reducing Project Related Uncertainty in the 'Fuzzy Front End' of Innovation: A Comparison of German and Japanese Product Innovation Projects", International Journal of Product Development, Vol.1, No.1, pp.43-65, 2004.

[5] J. Kim and D. Wilemon, "Focusing the Fuzzy Front End in New Product Development", R&D Management, Vol.32, No.4, pp.269-279, 2002.

[6] K. Ulrich and S. Eppinger, "Product Design and Development", 5th edition, p. 432, McGraw-Hill, 2011.

[7] R. Cooper, "Winning at New Products: Accelerating the Process from Idea to Launch", 4th edition, p. 408, Basic Books, USA, 2011.

[8] S. Conn, "NPD Success Factors: A Review of the Literature", Lappeenranta, Finland, Lappeenranta University of Technology, research report 160, p.34, 2005.

[9] M. Bastic, "Success Factors in Transition Countries", European Journal of Marketing, Vol.7, No.1, pp.65-79, 2004.

[10] L. Dwyer and R. Mellor, "New Product Process Activities and Project Outcomes", R&D Management, Vol.21, No.1, pp.31-42, 1991.

[11] A. Khurana and S. Rosenthal, "Integrating the Fuzzy Front End of New Product Development", Sloan Management Review, Vol.38, No.2, pp.103-120, 1997.

[12] P. Koen, G. Ajamian, R. Burkart, A. Clamen, J. Davidson, R. D'Amore, C. Elkins, K. Herald, M. Incorvia, A. Johnson, R. Karol, R. Seibert, A. Slavejkov and K. Wagner, "Providing Clarity and a Common Language to the Fuzzy Front End". Research Technology Management, Vol.44, No.2, pp.46-55, 2001.

[13] R. Cooper and E. Kleinschmidt, "New Products: The Key Factors in Success", p.52, USA, American Marketing Association, 1991.

[14] L. Cohen, P. Kamienski and R. Espino, "Gate System Focuses Industrial Basic Research", Research and Technology Management, Vol.41, No.4, pp.34-37, 1998.

[15] F. Langerak, E. Hultink and H. Robben, "The Role of Predevelopment Activities in the Relationship Between Market Orientation and Performance", R&D Management, Vol.34, No.3, pp.295-309, 2004.

[16] A. Griffin, "Drivers of NPD Success: The 1997 PDMA Report", Product Development and Management Association, Chicago, 1997.

[17] H. Kärkkäinen and K. Elfvengren, "Role of Careful Customer Need Assessment in Product Innovation Management - Empirical Analysis", International Journal of Production Economics, Vol. 80, No. 1, pp.85-103, 2002.

[18] K. Goffin, F. Lemke and U. Koners, "Identifying Hidden Needs: Creating Breakthrough Products", Palgrave Macmillan, p.256, 2010.

[19] G. Urban and J. Hauser, "Design and Marketing of New Products", Prentice-Hall, New-Jersey, 1993.

[20] B. King, "Better Designs in Half the Time", Goal/QPC, Methuen, MA, 1989.

[21] L. Cohen, "Quality Function Deployment: How to Make QDF Work for You", Reading, Massachusetts, USA: Addison-Wesley, 1995.

[22] P. Smith and D. Reinertsen, "Developing Products in Half the Time: New Rules, New Tools", Van Nostrand Reinhold, USA, p.298, 1998.

[23] A. Osterwalder and Y. Pigneur, "Business Model Generation: A Handbook for Visionaries, Game Changers, and Challengers", p.288, John Wiley and Sons, 2010.

[24] S. Mizuno and Y. Akao, "Quality Function Deployment: A Company Wide, Quality Approach", JUSE Press, 1978.

[25] G. Altshuller and R. Shapiro, "On the Psychology of Inventiveness", Voprosy psihologii, No.6 (in Russian: О психологии изобретательского творчества / Вопросы психологии), 1956.

[26] G. Altshuller, "And Suddenly the Inventor Appeared: TRIZ, the Theory of Inventive Problem Solving", Technical Innovation Center Inc., p.171, Worcester, MA, USA, 1996.

[27] R. Jugulum and M. Sefik, "Building a Rubust Manufacturing Strategy", Computers & Industrial Engineering, Vol. 35, No. 1–2, pp. 225–228, 1998.

[28] H. Yamashina, T. Ito and H. Kawada, "Innovative Product Development Process by Integrating QFD and TRIZ", International Journal of Production Research, Vol. 40, No. 5, pp.1031-1050, 2002.

[29] T. Sakao, "A QFD-Centred Design Methodology for Environmentally Conscious Product Design", International Journal of Production Research, Vol. 45, No. 18–19, pp. 4143–4162, Sep. 2007.

[30] H. Kobayashi, "A Systematic Approach to Eco-innovative Product Design Based on Life Cycle Planning", Advanced Engineering Informatics, Vol. 20, No. 2, pp. 113–125, Apr. 2006.

[31] C.-S. Lin, L.-S. Chen and C.-C. Hsu, "An Innovative Approach for RFID Product Functions Development", Expert Systems with Applications, Vol. 38, No. 12, pp. 15523–15533, Nov. 2011.

[32] C.H. Yeh, J.C. Huang and C.K. Yu, "Integration of Four-Phase QFD and TRIZ in Product R&D: A Notebook Case Study", Research in Engineering Design, Vol. 22, No. 3, pp.125-141, 2010.

[33] C.T. Su and C.S. Lin, "A Case Study on the Application of Fuzzy QFD in TRIZ for Service Quality Improvement", Quality & Quantity, Vol. 42, No. 5, pp.563-578, 2008.

[34] G. Altshuller, "40 Principles Extended Edition: TRIZ Keys to Technical Innovation", Technical Innovation Center Inc., p.143, Worcester, MA, USA, 2005.

APPENDIX 1. THE QFD-MATRIX

Relative Weight	Weight / Importance	Customer needs \ Technical features	Powerclass 100-200 W (35/35)	Max weight 2 kg	Impact resistance (stand)	Vibration resistance (stand)	Sealing IP-class IP44 (dust, water)	Toleranse for accleration and retardion (same as for the fan)	COP > 1, Coefficient of performance	Repairing by changing the whole module	Plug & Play installation
10,3	4,0	Replacement of the fan cooling with a more effective technology	⊖						⊖		○
7,7	3,0	Effective cooling without loosing a good efficiency					▲			○	
10,3	4,0	Low energy consumption	○						⊖		
12,8	5,0	Thermal control of the casing			▲	▲	▲				▲
12,8	5,0	Good efficiency	⊖						⊖		
5,1	2,0	Easy serviceability		▲	○	○				⊖	
12,8	5,0	Reliability			⊖	⊖	⊖	⊖		○	○
10,3	4,0	Light weight (should be light)		⊖				⊖			
5,1	2,0	Size of the cooler should be small	○	○						▲	
12,8	5,0	Easy installation	⊖						⊖		
		Difficulty (0=Easy to Accomplish, 5 =Extremely Difficult)	5	2	3	3	4	2	5	1	5
		Max Relationship Value in Column	9	9	9	9	9	9	9	9	3
		Weight / Importance	369,2	112,8	143,6	143,6	135,9	207,7	415,4	112,8	82,1
		Relative Weight	21,4	6,5	8,3	8,3	7,9	12,1	24,1	6,5	4,8

Legend		
⊖	Strong Relationship	9
○	Moderate Relationship	3
▲	Weak Relationship	1

Selecting an economic indicator for assisting theory of inventive problem solving

Mariia Kozlova, Leonid Chechurin, Nikolai Efimov-Soini
School of Business and Management
Lappeenranta University of Technology
Lappeenranta, Finland
mariia.kozlova@lut.fi

Abstract—The theory of inventive problem solving known as TRIZ is shown to be a powerful tool in new product development. Easy-to-follow principles navigate engineers towards inventive solutions in design concept elaboration. However, lack of economic consideration offered by TRIZ may lead to improper decisions in terms of cost efficiency. This paper aims to select a suitable economic indicator that can be integrated into TRIZ practices. We analyze features of TRIZ, review available profitability and cost indicators, and offer a suitable option, illustrating its applicability with a case study. A simple in use single indicator coinciding with TRIZ is shown to bring economic consideration to engineering decision-making enabling cost-efficient solutions.

Keywords—TRIZ; new product development; levelized cost

I. INTRODUCTION

Industry spends a considerable share of design time to developing concepts [1]. For intensification of this process, engineers resort to different approaches and among them to a powerful tool for creative ideation known as theory of inventive problem solving or TRIZ [2]. Generating ideas needs to be complemented with their selection, and here a variety of methods can be used [3]. However, in the real world utilization of those approaches is shown to be relatively low and limited to such multi-criteria decision-making methods, as analytical hierarchy process and Pugh's matrix [4]. Nevertheless, insufficiency of economic consideration in selection methods may lead to cost-inefficient decisions.

This paper aims to review existing profitability and cost indicators and to select the most suitable one for its integration with TRIZ. Our analysis is shaped by peculiarities of TRIZ, main of which is its function orientation. In accordance with the TRIZ concept of ideal final result, a good device is no device, but its function is performed [5]. For example, instead of thinking of product elaboration to perform a needed function, it encourages to think how the function can be performed itself involving existing resources in target timing. This focus on function, rather than a product, should be also reflected in an economic indicator to provide their logical integration.

The authors would like to acknowledge Fortum Foundation, and TEKES, the Finnish Funding Agency for Innovation, and its program FiDiPro for their support.

II. REVIEW OF SELECTED INDICATORS

Traditional capital budgeting appraisal includes assessing future project cash flows and discounting them to reflect the time value of money. Based on that several profitability indicators can be calculated. The most widely used among them is Net Present Value (NPV) [6, 7]. Its origins are dated to times of Karl Marx [8] and Irving Fisher [9]. It represents the sum of discounted project-related cash flows that can be formulated as follows:

$$NPV = \sum_{n=0}^{N} \frac{I + (Rev + O\&M)_n}{(1+d)^n} \qquad (1)$$

where

I is initial investment,

Rev is cash inflows,

$O\&M$ are operating and maintenance expenses,

d is discount rate,

n is period,

N is total lifetime.

Here cash inflows are essentially take the positive sign, whereas outflows – the negative. As the result, positive NPV signifies economic viability of a considered project.

Although NPV is commonly used for investment project appraisal, its inability to bring into comparison projects with different lifespans provoked engineers to adopt the Equivalent Annual Annuity (EAA) [10] in order to compare smaller capital investments e.g. into separate technical devices. EAA converts NPV into equivalent annual annuity payments that solves mentioned inefficiency of NPV:

$$EAA = NPV * CRF \qquad (2)$$

where *CRF* is a capital recovery factor calculated as

$$CRF = \frac{d(1+d)^N}{(1+d)^N - 1} \qquad (3)$$

Nevertheless, evaluation of separate technical devices embedded into a bigger system is often characterizes by complications in estimating the revenue side owing to their minor contribution to the overall process or system. Addressing this issue the Value Engineering concept suggests using the life-cycle cost (LCC) estimate that is analogous to NPV approach, but excludes the revenue part [11]. It represents all design-related discounted costs, including its manufacturing and installation costs, operating and maintenance expenditures and disposal costs if any:

$$LCC = \sum_{n=0}^{N} \frac{I + O\&M_n}{(1+d)^n} \qquad (4)$$

However, excluded cash inflows from the calculation lead to inability of LCC to capture functionality or, in other words, it fails in comparing alternatives with different production profiles ceteris paribus.

In an attempt to resolve this drawback, we resort to the energy sector, where evaluation and comparison of technologies with different electricity production performance is an essential analytical necessity. So-called levelized cost of energy is widely used in business [12-14] and academia [15-19] as a single indicator of all technology-related costs weighted per a unit of expected electricity production by this technology over its lifetime [20]. Here we generalize it to any function referring to it as levelized function cost (LFC):

$$LFC = \frac{\sum_{n=0}^{N} \frac{(I+O\&M)_n}{(1+d)^n}}{\sum_{n=0}^{N} \frac{Q_n}{(1+d)^n}} \qquad (5)$$

where

Q is the function output in period n.

If annual output and O&M costs are constant over time, LFC formulation can be converted into annuity terms simplifying the calculation [20]:

$$LFC = \frac{I * CRF + O\&M_a}{Q_a} \qquad (6)$$

where

$O\&M_a$ is annual operating and maintenance costs and

Q_a is annual function output.

Summarizing advantages and disadvantages of presented indicators we arrive at the following picture:

TABLE I. COMPARISON OF INDICATORS

Feature	NPV	EAA	LCC	LFC
Comparable for different lifetime	✗	✓	✗	✓
No need to estimate revenues	✗	✗	✓	✓
Comparable for different productivity	✓	✓	✗	✓

To outline, both NPV and EAA require revenue estimation, LCC concentrates on costs only, but fails to incorporate productivity difference, whereas LFC with no need to estimate inflows reflects both, productivity and durability of estimated projects or design concepts. Therefore, it is chosen as the most suitable for TRIZ needs and peculiarities. Further, we illustrate its application on a numerical case study.

III. CASE ILLUSTRATION

The illustration is built on a real world case of improving a hydraulic press for fixing rings in the flange. The technology finds its usage in a variety of repairing and maintenance tasks in automotive industry. Disadvantages of the conventional hydraulic press include low durability and the need for specialized workers for its operation. A new proposed tool with pneumatic actuator requires a less skilled (and lower paid) worker, serves for longer period, performs the same operation faster, but has substantially higher initial costs. The specifications of both designs are presented in the Table 2.

TABLE II. TECHNOLOGY SPECIFICATION

Feature	Old design	New design
Lifetime, (years)	3	10
Time for one operation (min)	1,0	0,8
Manufacturing cost (rub.)	20 000	170 000
Required worker salary (rub./month)	40 000	22 000
Maintenance cost (rub./year)	200	500

The essential question is whether to opt for the improved device. From the engineering perspective both designs perform the same function and only economic consideration can provide an answer with respect to cost efficiency of this decision.

Although valuation of these alternatives can be addressed with all discussed indicators, our analysis shows that NPV and EAA estimations are sensitive to defining the price level for the operation and thus, allowing manipulation with results. Life-cycle cost estimate shows already obvious fact that the new design is more expensive. Whereas only the levelized function cost provides unbiased and useful insight for decision-making.

Our calculations show that the cost of the same operation performed with the new device is more than twice less than with the old one (1,97 and 4,12 rubles per operation respectively with discount rate 10%).

All considered indicators, including recommended LFC are represented by crisp numbers, thus if used standalone, are not capable to capture uncertainty and impression of input estimates. Therefore, for deeper analysis we encourage using such techniques as sensitivity analysis, simulation, fuzzy set based tools etc. In addition, LFC could be used as one of criteria in multi-criteria decision-making models.

IV. CONCLUSION

This paper proposes the levelized function cost for integration with TRIZ to supplement ideation with economically wise selection of ideas. LFC expresses a cost of a unit of function produced by a product. This function orientation makes it an ideal candidate for integration with TRIZ. The numerical example demonstrates its simplicity and rationality. Reflecting all life-cycle costs of a product, its durability and productivity in a single indicator, LFC becomes relevant and appealing for engineers. It can be easily adopted into engineering practice by either stand-alone use or integrated to specialized software.

ACKNOWLEDGMENT

We would like to acknowledge the input of Prof. Mikael Collan from LUT School of Business and Management. He provoked and inspired the authors to bridge design and evaluation methods.

REFERENCES

[1] G. Ullman David, "The Mechanical Design Process [Bok]," Oregon State: McGraw-Hill, 2010.

[2] G.S. Altshuller and R.B. Shapiro, "Psychology of inventive creativity," Issues of Psychology, vol. 6, pp. 37-49, 1956.

[3] G.E. Okudan and S. Tauhid, "Concept selection methods-a literature review from 1980 to 2008," International Journal of Design Engineering, vol. 1, pp. 243-277, 2008.

[4] M. Salonen and M. Perttula, "Utilization of concept selection methods: a survey of Finnish industry," in ASME 2005 International Design Engineering Technical Conferences and Computers and Information in Engineering Conference, pp. 527-535, 2005.

[5] L. Chechurin, "TRIZ in science. Reviewing indexed publications." in TRIZ future conference, 2015.

[6] P.A. Ryan and G.P. Ryan, "Capital budgeting practices of the Fortune 1000: how have things changed," Journal of Business and Management, vol. 8, pp. 355-364, 2002.

[7] J.R. Graham and C.R. Harvey, "The theory and practice of corporate finance: evidence from the field," J.Financ.Econ., vol. 60, pp. 187-243, 5. 2001.

[8] K. Marx, Capital: A Critique of Political Economy, Vol. III. The Process of Capitalist Production as a Whole, Chicago: Charles H. Kerr and Co., 1894, .

[9] I. Fisher, The Rate of Interest: Its nature, determination and relation to economic phenomena, Macmillan, 1907, .

[10] T.W. Jones and J.D. Smith, "An historical perspective of net present value and equivalent annual cost," The Accounting Historians Journal, pp. 103-110, 1982.

[11] D. Younker, Value engineering: analysis and methodology, CRC Press, 2003, .

[12] US Energy Information Administration, "Levelized Cost and Levelized Avoided Cost of New Generation Resources in the Annual Energy Outlook 2015," vol. 2015, 2015.

[13] World Energy Council, "World Energy Perspective. Cost of Energy Technologies," vol. 2015, 2013.

[14] Bloomberg New Energy Finance, "The cost landscape of solar and wind," vol. 2015, 2015.

[15] K. Branker, M. Pathak and J.M. Pearce, "A review of solar photovoltaic levelized cost of electricity," Renewable and Sustainable Energy Reviews, vol. 15, pp. 4470-4482, 2011.

[16] M. Campbell, P. Aschenbrenner, J. Blunden, E. Smeloff and S. Wright, "The drivers of the levelized cost of electricity for utility-scale photovoltaics," White Paper: SunPower Corporation, 2008.

[17] C. Breyer and A. Gerlach, "Global overview on grid parity," Prog Photovoltaics Res Appl, vol. 21, pp. 121-136, 2013.

[18] X. Ouyang and B. Lin, "Levelized cost of electricity (LCOE) of renewable energies and required subsidies in China," Energy Policy, vol. 70, pp. 64-73, 2014.

[19] J. Hernández-Moro and J. Martínez-Duart, "Analytical model for solar PV and CSP electricity costs: Present LCOE values and their future evolution," Renewable and Sustainable Energy Reviews, vol. 20, pp. 119-132, 2013.

[20] W. Short, D.J. Packey and T. Holt, A manual for the economic evaluation of energy efficiency and renewable energy technologies, University Press of the Pacific, 2005, .